Converged Security Metrics

A Top 25 Set of Solutions

2023 Edition

Jim McConnell

Converged Security Metrics – 2023 Edition

For more information, write to Jim McConnell at:

top25@top25securitymetrics.com

Or visit our website at: top25securitymetrics.com

ISBN – 979-8-9869648-0-5 Paperback

ISBN – 979-8-9869648-1-2 eBook

ISBN – 979-8-9869648-2-9 Hard Cover

Registered with U.S. Library of Congress

Author: Jim McConnell
Editor: Maria Alicia Herrera
Book Cover Designer: Rob Williams

Honoring the Honorable

In my opinion, no book came from a single person's wisdom, knowledge, or experiences, now that wonderful book, called *The Bible*, truly singularly inspired. I humbly bring one person's perspective from countless people that have granted me access to their wisdom, knowledge, and experience over the last 30 years of doing something related to security. My first exposure was in 1984 through the first mentor that I recognized as a modern-day mentor; he later became one of my supervisors.

I first want to honor my Heavenly Father and My Lord Jesus. Your grace and voice have saved me and brought me through a massive, interconnected journey to this point. I am nothing without what you have done in my life. I surrender this book to your use.

Second to honor is my fantastic wife of over 30 years, Diane, my gifted son, James Jr., my mom and my late father, my sister Lynn, and my brothers Bob and Sam. Thank you all for your patience and never-ending grace. Thank you for being more family to me than I deserve. I love you all!!

To pastors in my life, particularly the late Pastor Dale Brooks, his wife Mama Kaye Brooks and Pastor Robert and Debbie Morris, you, and countless peers, have been second earthly fathers and mothers. May I fulfill *1 Timothy 5:17* on each of you before I leave this earth.

To my law enforcement, military, first responder friends, team members, and mentees, I am so grateful for your service and sacrifice for all of us. May we never forget.

To my industry peers … this is when networking meets the power of friendship. You all have impacted me and taught me in so many immeasurable ways. The list is in the thousands of stories I could talk about what I have learned from all of you. I genuinely wish there were a way to publicly honor you, just know that I am also writing this book for you, your organizations, and your communities, as my way to give back a little of what I have been given.

To friends, teachers, and people I have never met that have prayed for me. Thank you for the laughs, the fun, the grace, and the wisdom you have all bestowed upon me.

Table of Contents

Introduction

I have had the privilege of being involved with security as a significant or full-time part of my job for about 30 years. I am so overwhelmed by the blessings, grace, wisdom, and knowledge I have been given by thousands (if not tens of thousands) of amazing people and professionals. It has been said that the only person that can call you or me an *expert* is a judge. Since I do not like being in front of a judge, I will stick to just being a blessed security professional wanting to bless others. When it comes to security metrics, people like Caroline Wong, author of *Security Metrics, a Beginner's Guide*, Jon Boylan, and the entire team at National Institute of Standards and Technology (NIST) and the Data Nerds at Tableau, I can only wish to be in the "second chair" to their work and others that have trailblazed this much longer and better than me.

My professional friend, Susan Menaker, asked me and many others in a profound and emotional talk, "To Care," about securing the assets we were charged with securing. Susan, I care! So, this journey is about improving the security of everyone and everything that gives us freedom, hope, joy, love, compassion, and the desire to pay our God-given wisdom, knowledge, and experience forward to others. This is part of my "why" for writing this book. Thank you, Susan!

I am good at listening to security problems, which often are tied to excuses of what we cannot solve to meet a security challenge, whether it is business, people, process, or technical. I love what one of my former Chief Security Officers (CSOs) used to say, "Don't bring me a solution looking for a problem." Security in any organization does not lack problems; maybe this book will be one of the tools in the toolbox for some of those problems or support the justification for other solutions.

I am also enthusiastic about understanding how certain words are used, the audience's definitions, context, and scope. Hence, you will see discussions about the scope and context of terminology throughout this book.

My passion for this book is that it improves the security of our free nations, companies (small and large), entrepreneurs, non-profits, faith-based community, and us as citizens of Earth, for the better. But also ends a ton of excuses about improving and maturing security of *stuff*, people, technology, processes, and organizations.

This is *A* tool, not *THE* tool. Each function and each metric under each function are meant to challenge the analyst, engineer, executives, suppliers, customers, the BoD, the Secretary of a government agency, and even Congress/Parliament. What if these metrics, or even a subset, were part of a poster-size exhibit in every BoD and congressional committee meeting on security?

Countless times, I have been challenged by corporate leadership and my own fears, uncertainties, and doubts, with how to measure what I do in security, and in many ways, I have used many excuses … but that is for a fun chapter later.

There are many resources on this topic, from Caroline Wong's terrific book on security metrics to the fantastic work I have seen at NIST.

Caroline's Resource

My first real metrics work that I *owned* really started when I started doing BS7799/ISO17799[1] converged security assessments with my friend Israel Vila in 2000. We used a little piece of software out of the United Kingdom called COBRA by C&A Systems that measured, with some consistency, BS7799 compliance across all ten security domains. When we would conduct each assessment, we could compare the metrics between all the clients. However, it was not necessarily my *ah-ha* moment. I have seen, analyzed,

[1] A standard is published in two parts: (Part 1) ISO/IEC 17799: Code of practice for information security management; (Part 2) BS 7799: Information security management – specifications with guidance for use.

challenged, and even owned more statistics and metrics in the fraud area of security than anything else. Even in a place where return on investment (ROI) should be easy, like fraud, it was always the most difficult to defend.

My *ah-ha* moment to get this first edition of this book published, came in two parts; one was a bunch of security manufacturers claiming (in 2018) that "there are no cyber security standards for security manufacturers," and two, a senior CSO said on a panel I was on, in mid-2022, "there are no metrics for physical security."

One of the more challenging areas that I and many others in the industry (some with over 30 years' experience) have discussed has been connecting various security functions to the business. One of the many ways to connect to the business is to measure and mature security, like the business. With experience, research, and industry-leader input, I developed a collection of Top 25 metrics that can be evaluated to assess a security team, a security function, and/or an organization's security. These metrics can also help understand the organization's belief system about security with measurable facts along with other value-added use cases. I hope that much more intelligent people will introduce these or likely better metrics in more formal documented programs under great organizations such as ASIS, ATAP, BEPP, IFPO, IPSB, ISACA, ISSA, NACD,

NIST, SIA, etc.[2] Still, until then, I and many others believe there is a need for a starting point or a least a 2023 update to how security metrics have been thought about in the past.

So this little book is my attempt to: one, give back a tiny slice of what I have been given; two, provide a solution that is broad and deep at the same time; three, eliminate countless metrics excuses when you or I hear them in the future; four, improve security anywhere the principles in this book are applied; and five, make a lot of security professionals and audiences of security metrics very uncomfortable. Why? So, they/we take action. Metrics Rule #1: Ensure your audience is ready for the answer.

The existence and the implementation of these metrics can trigger difficult conversations within an organization and may have a negative and/or high consequence impact on individuals, from engineers to senior leadership. I also am very aware and have been on the wrong end of this aspect of metrics. BoD/Judge/Jury/Regulator: "Hey CEO/CSO/CISO, what do you mean we are not at 100% across the enterprise on [*pick the metrics area*]."

---

[2] American Society for Industrial Security (ASIS), Association of Threat Assessment Professionals (ATAP), Board of Executive Protection Professionals (BEPP), International Foundation for Protection Officers (IFPO), International Protective Security Board (IPSB), Information Systems Audit and Control Association (ISACA), Information Systems Security Association (ISSA), National Association of Corporate Directors (NACD), National Institute of Standards and Technology (NIST), Security Industry Association (SIA)

This book is for the seasoned security professional, the new security professional wanting to learn, but also for the *customer* of these metrics such as the executives or congress member.

Let us get started. I hope this gets you thinking and provides some real ideas to improve the security for which you are responsible. Thank you for time in pursuing this resource.

Purpose and ROI for Security Metrics

Security metrics should be used by organizations that have or need security. For example, if you need locks on your door, you need security. A small business can have 35 suppliers in its supply chain within 30 days of starting the company.

The set metrics in this book are *A* tool, not *THE* tool, which can help an organization assess the maturity of their various security functions. Suppose you do not think you have a security program or a specific security function to assess. In that case, you might have some general security maturity issues, not just metrics issues in these areas.

Some great questions I recently received from one of the most giving executive protection professionals and instructors I have met, Joe LaSorsa from LaSorsa and Associates, really helped me think about this from a wider perspective. I have adjusted my discussion with Joe to cover a broader security discipline not just executive protection.

Question 1:

What drives tracking security key performance indicators (KPIs) – To show an ROI, customer satisfaction, proof of work, tracking effectiveness, all the above, or other?

My thoughts:

All the above, and in addition, it is also driven by team awareness, leadership awareness, budget, gear, training needs, audits, litigation, and maturity goals.

Question 2:

Metrics versus statistics – How do you use stats/data to supply meaningful metrics in the form of KPIs? Are they directed from above or self-served? Why did you choose those specific topics to track as KPIs?

My thoughts:

Team members or technology should supply the inputs of the stats, see Supplier, Inputs, Outputs, and Customers (SIPOC), the analytics work, reporting work, and dashboard work, hopefully, turns the stats into KPIs. Some KPIs rarely get reported beyond the managers/team leaders, however some end up with the Board of Directors (BoD) and even the Client.

Question 3:

Who is interested in KPIs? Who are your KPIs presented to or evaluated by? Clients, executives, team members, etc.?

My thoughts:

Interest in KPIs comes in waves – team members show interest when they want vacation or question workload (not necessarily for self-improvement). KPI requests from the Program Chief and CSO can be monthly, or ad hoc (for example, as part of questioning by senior leaders/finance). Clients may be presented KPIs or request KPIs due to contract performance or compliance requirements. Some audiences may even require metrics in near real-time, so consider that need in your design of your metrics program.

Question 4:

How do KPIs demonstrate an ROI for your security program?

My thoughts:

I think ROI in the Security world should be based on:

1. Actual activities like negative event (late, violence, embarrassment, etc.)
2. Soft ROI: customer satisfaction or net promoter score (NPS), perceived increase in productivity. Sometimes, like in the executive protection function, for example, we should measure based on comments such as, "how did you guys get me here so fast with all that traffic." Tracking of comments (good and bad) is sometimes the best way to generate qualitative measures.

Question 5:

Have KPIs influenced your budget?

My thoughts:

The lack of KPIs – yes; the existence of KPIs – yes, but only certain ones.

Question 6:

Do you use KPIs to prove the negative in your security program? If so, how have you done this effectively? (*Proving the negative means showing that your security/risk management efforts have led to decreased risk or have reduced the number of incidents, etc.*)

My thoughts:

KPIs are rarely used to prove the negative; but you need to be able to defend them. Clients and the business will likely forget the next time there is an incident. Instead, we need to have stories and some stats, if needed.

Question 7:

Do you have general comments about the use of KPIs in security?

My thoughts:

Get tools, spreadsheets, anything necessary in place to start collecting stats and stories/comments, now, so all parties get to the point that metrics are BAU/SOP. Dedicate a person that you train in Analytics, Visualization, and Reporting. Realize there are different audiences for the output, hence the output should not be one size fits all.

"Who owns Security Metrics?"

… asks your CEO, BoD, or Agency Secretary. As some of you know, I dig a lot into the words used, their scope, and their context. Therefore, during three mentoring sessions recently, I was challenged with this dilemma. Are senior security titles deceiving our organization's leadership, non-maliciously, of course? Does our leadership, regulators, customers, etc., think that the "Chief [of something related to security] Officer" covers EVERYTHING in that area? If we do not tell them what our scope DOESN'T cover, do they assume we cover ALL? To help, here is a possible responsibility assignment (RACI) chart to consider, if ownership is a challenge in your organization.

Possible Security RACI Chart

Typical Title	Classic Acronym	General Responsibility (*ALL* in every language means ALL – All business units, all countries, all legal entities, all assets, all employees, all suppliers, all vendors, all customers ... you know, *ALL*)	Operate (Do the technical/ operational work of items in red font)	Manage (Ensure the humans that do the technical/ operational work, actually complete the work, but they do not need to report to you)	Govern (Provide policies, standards, guidelines, "oversight" for items in **bold** font, but do not DO or MANAGE)	Metrics (Collect metrics from all functional owners, analyze metrics, and report them to various audiences)
Chief **Security** Officer	CSO	ALL Areas of **Security** (cyber, info, physical, investigations, fraud, law enforcement, personnel security, etc.) for all assets				
Chief **Cyber Security** Officer	CCSO	ALL **Cyber Security** of ALL electronic assets (not all information security areas)				
Chief **Information Security** Officer	CISO	ALL **Information Security** for ALL organization's information assets. ALL Cyber and Physical Protections. ALL Confidentiality, Integrity, and Availability (CIA) elements/controls				
Chief **Privacy** Officer	CPO	ALL elements of **Privacy** (including ALL security and legal elements of private information/data) – employee, customer, supplier, other's data/information, for ALL jurisdictions				
Chief **Compliance** Officer	CCO	ALL **Compliance** for ALL laws, regulations, customer contracts, supplier contracts, for all elements of the organization				
Chief **Risk** Officer	CRO	ALL types of **risks** (man-made, third-party, natural, financial, etc.)				
Chief **Audit** Officer/Executive	CAO/CAE	ALL types of **audits** (financial, security, customer, supply chain, government, etc.)				
Chief **Physical Security** Officer	CPSO	ALL **Physical Security** for ALL locations housing company assets				
Head of **Investigations**		ALL **investigations** (employee, customer, fraud, supply chain)				
Or am I really ... ?						
Chief Subset of ________ Officer		Only **some of the** functions from an area described above				

How to Use this Book

Simple, use this book as a tool on your way to implementing changes to your organization's security functions.

Step 1: Is your audience ready for the results? No? Get them ready! Measuring is not usually the problem; it is the audience's preparedness for the results.

Step 2: Use "The Golf Course Solution" – here is the scenario: three of your executives are on the Golf Course; one has a bad report from your metrics implementation; one has a good report. How would that change the competition on the course and back in the office? This is not about *naming and shaming*, but it is about ownership by a human's name – accountability.

Step 3: Do you have an excellent, agreed-upon set of definitions for security, the scope of security, and the elements of your metrics program? If not, start there.

You do not want a metrics program to be flawed by disagreement on definitions/scope/context? In the Appendix (see *Definitions*), I supply some definitions for you to consider in the context of a robust metrics program.

My experience in providing a list of questions runs into long discussions about context and scope and so much back-and-forth that we can get to a measurable way forward for ongoing improvements. I believe percentage-based reporting provides some immediate solutions but is never a replacement for healthy question and answer sessions.

To get us started with this first tool, I provide, as one of many available solutions, a yearly *Top 25* set of metrics modeled after the many Top ## tools available in various industries, including in the security world. This was not meant to gather 1001 opinions on what should be the *Top* but to start with a 2023 edition to improve to a better *Top 25* in future years.

I genuinely believe this book and its embedded *tools* can trigger a broader and sometimes complex conversation about improving an organization's security program, individual function. Hopefully that conversation includes discussions about the metrics listed, such as which to use, which ones are not in the book, and even which ones are unique to your organization or industry.

It is simple to use and covers a broad set of metrics, but it is purposefully challenging. Run it against a sample of your organization or your entire organization or require it of your suppliers; the key is to start somewhere. Best to start with what is in the mirror.

I recommend you take each chapter/function, review it, adapt it for your organization, then look at the table, add a column with the human owner or human owner by business unit, and maybe a column for the system(s) of record, that would be the point of contact to gather the denominator and numerator for each of the *Top 25* metrics.

Then start. Bring these owners together, set a deadline, set a collection mechanism, then see where you end up with your first round. Use the results to communicate where you and the organization are and provide solutions to "clear the red" for your senior leadership. Insert them in your supply chain contracts and performance objectives. Auditors ask for these in your audits. Key is "strive to improve."

I recognize that some organizations are not at the maturity level in their security management program to start measuring it … but please start.

Each functional set of metrics focuses on a different part of a security capability that may or may not exist within your organization. It does not mean the security capability you are measuring is reporting to a department with the word *security*, *risk*, *compliance*, or *fraud* in its name. This is about measuring a FUNCTION. Suppose you have one of these functions in a separate group or managed differently by a business unit/legal entity. In that case, it is an exciting time to measure them using these metrics and compare the output for potential improvements or synergies (there, the finance people will be happy now that I used that word). You may even outsource the function or have decided that the function is not important, does not mean the metrics are not needed. Finally, you can use these or similar metrics to measure the security of your supply chain participants. Also, you might find a specific metric irrelevant to your organization, form a replacement, and replace it with a better one. Just start.

You, the owner of your metrics program and the overall organization, also need to recognize the hard part and get chartered and buy-in so you have a high integrity, defendable metrics program including high confidence numerators and denominators.

With a couple of obvious exceptions, all the topics/chapters of metrics are set up in the same format for consistency and ease of reporting. This table below is an example of that format.

	The Metric	Best Score
1	Percentage of employees with fun colorful bricks on their desk to play with	100%
2	Percentage of desks with dog beds	100%
3	Percentage of office skateboards with recent maintenance	100%

Column 1 is simply a number; you should not use it for prioritization or any other purposeful order, but if you contact

me, I might ask you for the number, so I have context to your questions.

Column 2 is the basic description or formula of the metric. The key is ensuring the audience agrees on the metrics *AND* the scope of the words that make up the metrics.

Column 3 is a simple representation of what would naturally be the BEST score. Notice, I did not say *best practice* (my apologies to the lawyers), *normal*, *average*, *minimum*, *industry standard*, or *common*, or even *KPI or customer performance indicator (CPI)*. This is about communicating where the organization is TODAY, and tomorrow, and …; one percent compliant is a result that the organization may have never measured before. You have started.

Reality, Not Naïve, But Jim, Caveats, What Abouts

This is not to avoid constructive feedback on this book and the metrics, but I wanted to address arguments I have had with myself (okay, and others) about metrics and this tool.

- Now, I am not naive to not expect one of your customers, BoD, or even the government to come in with this tool or revision of it, to check their supplier (you, in this case) or how your organization is governing its security and that it could be a significant challenge for you to respond comprehensively.
- This might not work in a government-owned enterprise.
- I know I will get negative feedback from not getting widespread input or following some standards-based metrics approach. That is okay; I wanted to provide a functional solution to "move the needle."
- *But Jim, you just do not know my company, industry, environment, boss, the politics, etc.* You're right, I do not and no where do I claim to. However, I wrote this to where my amazing son's small medium business (SMB) or a church or the largest company in the world, could use some or all of these metrics. As the late Carman said so well in one of his songs, "From the

Church House to the White House" – this phrase encapsulates my vision for this book/tool.

- *Jim, I disagree that this metric would improve security.* I respect that, contact me with your suggested change for the next edition. I suspect there are thousands of metrics that could improve security.
- *Jim, these are going to make me look bad.* I am sorry about that but work with your leadership about the pros and cons of an overall metrics program first. Second, challenge yourself with integrity of process and transparency on the ones that you need to step up and own.
- *Jim, why didn't you include XYZ?* In this first book, I had to stop somewhere, otherwise I would never get it published out of my desire to provide more solutions. Look at the 2024 list (towards the end of the book) of just some of the functional areas about which I am already thinking. After my next book is completed in early 2023, I will jump on working on the 2024 edition. But please, contact me with your ideas!!
- I am aware that some people will put some or all security functions of a converged security program under the heading of "physical security" or other organizational titles; some even say there are only two organizational models – cyber and physical. I do not think this way, sorry, just me … and thus, for this topic also, I am not limiting coverage.

- I have realized the vastness of cyber, application, network, privacy security areas, and the amount of disagreement over what it means to be a "mature" security program in these areas. I decided to put any *Top 25* attempt for these areas on the back burner and focus on areas that can help most of a converged security program and most organizations, regardless of size, type, ownership model, with a few cyber security metrics included, in context. But I haven't forgotten my cyber colleagues, so stay tuned.
- Disclaimers:
  - Vendors listed are not endorsements, just ones I have enjoyed the journey with.
  - If there is a link to a product in the book or website, it may be an affiliate link.
  - My contributions to the Telecommunications Industry Association (TIA) were fun and I appreciate their help and approval with my supply chain updates.
  - My normal disclaimers related to my employer, employment, and compliant to their Code of Conduct are located at https://askmcconnell.com/disclaimers. I have tried very hard to comply with the wonderful guidance and guardrails from my employer's legal and ethics team.

A Tool for the Tool

"Red is Bad, Green is Good" … is a saying from the old-style network operations centers (NOCs).

When I first imagined (okay, like crazy dreaming) the output of this tool in front of a BoD or even the crazier idea of a government oversight committee, I envisioned a large poster board with twenty-five red and green pie charts.

… Simple, but effective, to draw critical audiences to ask blunt questions about … the red.

Multiple poster boards for comparing department to department or agency to agency would get me even more excited about how many excuses can get squashed and we can genuinely scale security improvements.

You do not need to use (just) pie charts or even red and green. I only use them to show how a simple visualization can make an impact. I know there is even controversy about using these for visualization; as simple as these options are, I beg you, just visualize your metrics results. The important part is still … start … where you and your audience can feel a powerful sense of value, integrity, and assurance in what you see and learn about your organization's security in an auditable and defended way.

Automation will come as these metrics should not be hard to collect.

A few years ago, I became a fan of Tableau (compared to Excel or Numbers) for my daily analytics work. The tool is not as critical as knowing how to use it to bring value to your audience, so regardless of whether you use Excel or Numbers or some standard tool in your organization, go for it. I am showing you the metrics in Tableau only to give you another example of a solution in case you get more excuses for implementing these metrics.

Sample Metrics Collection Template

	A	B	D	E	F
1	Agency	Function	Measure Name	Total Elements (Denominator)	Number of Compliant Elements (Numerator)
2	Grocery Chain	InfoSec	InfoSec-1	95478	4774
3	Grocery Chain	InfoSec	InfoSec-2	41553	831
4	Grocery Chain	InfoSec	InfoSec-3	39025	25756
5	Grocery Chain	InfoSec	InfoSec-4	7033	4923
6	Grocery Chain	InfoSec	InfoSec-5	79947	1599
7	Grocery Chain	InfoSec	InfoSec-6	25241	8582
8	Grocery Chain	InfoSec	InfoSec-7	25012	2501
9	Grocery Chain	InfoSec	InfoSec-8	37256	7451
10	Grocery Chain	InfoSec	InfoSec-9	7654	842
11	Grocery Chain	InfoSec	InfoSec-10	66515	20620
12	Grocery Chain	InfoSec	InfoSec-11	62802	51497
13	Grocery Chain	InfoSec	InfoSec-12	1652	1140
14	Grocery Chain	InfoSec	InfoSec-13	78460	53353
15	Grocery Chain	InfoSec	InfoSec-14	63128	9469
16	Grocery Chain	InfoSec	InfoSec-15	6907	3661
17	Grocery Chain	InfoSec	InfoSec-16	92266	70122
18	Grocery Chain	InfoSec	InfoSec-17	75260	49671
19	Grocery Chain	InfoSec	InfoSec-18	24959	22962
20	Grocery Chain	InfoSec	InfoSec-19	95007	63655
21	Grocery Chain	InfoSec	InfoSec-20	26482	18802
22	Grocery Chain	InfoSec	InfoSec-21	72204	30325
23	Grocery Chain	InfoSec	InfoSec-22	7783	3814
24	Grocery Chain	InfoSec	InfoSec-23	44441	18221
25	Grocery Chain	InfoSec	InfoSec-24	13585	3940
26	Grocery Chain	InfoSec	InfoSec-25	79448	17479
27	Grocery Chain	PhySec	PhySec-1	22342	8713
28	Grocery Chain	PhySec	PhySec-2	47628	19051

Sample Metrics Dashboard

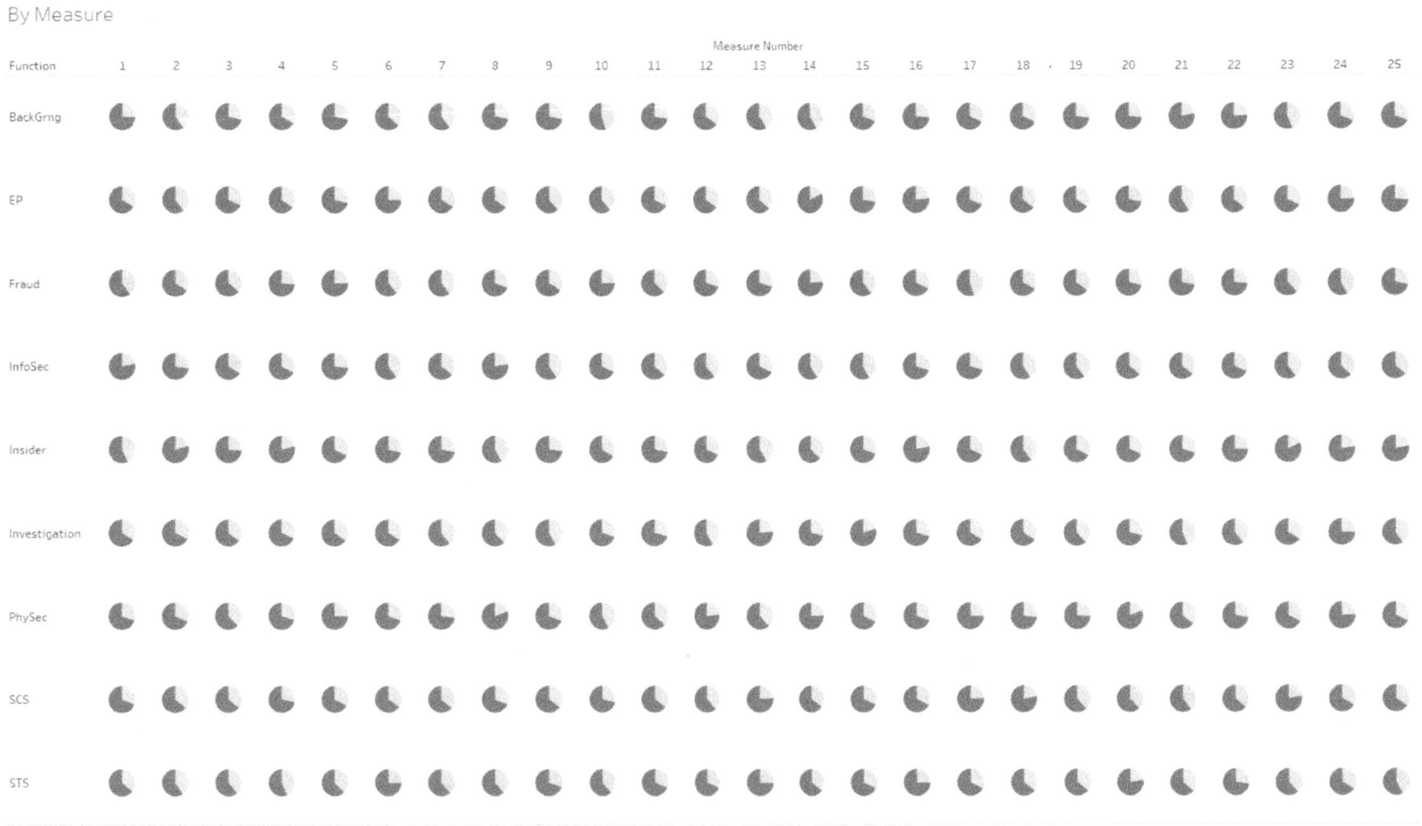

A Tool vs. The Tool

One of the critical pieces of wisdom I have tried to pass along more than anything else is:

1. *All* in every language means All – This came from one of the very impactful pastors in my life, Pastor Robert Morris. So, when implementing metrics, always remember to implement them to ALL the company, organization, enterprise, legal entities, business units, countries, etc. Your audience believes any metrics reporting includes ALL, so if you do not include a scope statement of what you are and are not included, you and they must assume ALL is covered. All business units, all counties, all legal entities, all assets, all employees, all suppliers, all vendors, all customers, you know ALL.
2. "A" vs. "The"
   a. "A" perspective/tool means there can be alternatives, some worse, some better vs. "The" perspective/tool might lead someone to believe there is only ONE perspective or tool in the security and metrics world, which is never the case.

These metrics, at a minimum, should be a conversation starter inside your organization. The first goal is that this book attempts to end the (primary) excuses that metrics do not exist in the public arena.

"But I/we don't like *that* specific metric!" Check yourself, please. Is that because it makes you/your function/or the company look bad, or it is hard to achieve or measure, or is it because it is truly not relevant to your organization or industry? Remember, just because you do not do a function does not mean you should not be, thus measuring its maturity. When I talk to pastors about church security, they are surprised at how many security functions come under their leadership; that is part of *church security* (not just metrics). But yes, if you find a metric that honestly does not fit, replace it, and find one that is better, more challenging, more robust.

When I started working with Tableau, one of my first lessons learned was how deceptive a simple pie chart could be. The primary issue is that humans have a tough time judging the size/context of an angle or slice of the graph. The metrics in this book do not end that concern, but it does try to minimize the pie chart challenge. I recommend using a pie chart, as it is a simple two-color visual where the angles of the slices are not as critical. Remember, Red is Bad, Green is Good. A more sophisticated delivery and analytics method through dashboards should be in your future.

Same with Excel vs. Numbers vs. Tableau vs. other reporting tools – They are all *A* tool but not *The* tool. As you move from manual information or data collection for this metrics program to a more data feed or automated program maturity, Excel and Numbers might limit you more than a tool such as Tableau.

This is a 2023 edition, aka *1.0*, if you are just getting started. If this is *Level 0* on your capability maturity model staircase graphic, when it comes to measuring your security, it is a starting point. Year 2023 and beyond will come; we all can improve these metrics and thus your security together. More on contributing to the 2024 edition is at the end of the book.

… But please stop making excuses … it will catch up with you … including your boss/BoD/oversight committee finding this book before you do …

Why 25?

My first topic was supply chain security about a year ago, and I started with a simple *Top 10*, then I realized:

- How many areas was I missing and prioritizing other metrics to fit into a Top 10?
- It is not about whether it is 25 or 125; it is a tool, solution, and excuse-slayer. How many hammers did my dad have in his toolbox when I was growing up?
- How I felt the need for more options for my audience (you) to consider.
- How many more areas of vulnerability and threat should we cover in this first edition.
- How many times *Top 25* are in other Top XX security and non-security lists in the industry.

Yes, I know for one of them, I got to a Top 30, but that is for a longer chapter or a taller glass of Sweet Iced Tea.

Remember, the purpose here is not about 25 or implementing all 25 today across all functions. We have over 500 metrics in this book, and even I'm amazed at what I still haven't covered. This is about providing you with some ideas, options, and excuse-slayers across a wide range (not all) of security disciplines to improve your organization's security and thus improve the security of society overall.

Why Percentages?

You will see I made all these into a percentage to make it consistent and hopefully easier as a starting point. It will also likely challenge an immature *AND* maturing program to figure out how to define, collect, report, analyze and present both high integrity denominators and numerators that drive your operation's maturity through percentages. This is ***not*** the only way of measuring your security, implementing metrics, KPIs, or CPIs; it is just *one* of many ways ... ask NIST!

Also, you have many ideas and tools that measure security more statistically (e.g., 100 alarms per day, 1,000 spam attempts per month) that you should still measure, analyze, trend, report, ***and*** implement solutions to improve these stats. The metrics in this book are not a replacement for statistic-based measuring of other security and non-security related *stuff*, maturity, or risk.

For some metrics, a result of 0% is the best; for other metrics, 100% is best, sometimes a mix/balance. Tracking over time with a dashboard should be your goal. Adjust for your organization, clients, culture, and message but also measure with 100% integrity. Obviously, I and this book can't answer for the audience if 100% or 80% or even 10% is *good enough*. Checklists that offer only a *Yes* or *No* answer is the same as saying the only metric is 0% or 100% and that won't mature a program. Also, be careful that 0% or 100% is focused on maturity and not scope of coverage.

This is overcomplicating the concept here, but yes, I agree that if solving 10% of a metric reduces your vulnerability by 99% and going from 11% to 100% only gets you to 99.999% (Five 9's) … move on. But never forget the old security saying, a bad person only needs to exploit one vulnerability. The key is not to believe your environment will stay at 99.999% forever.

You also might find that for some optics or political or financial reasons, you can only implement one of the metrics from a particular chapter/function; again, that is a start, do it with excellence. But please do not forget about the other 24; they are not useless, even though your organization may have an excuse or valid risk reason to "put them on the back burner."

Validation: Trust but Verify: As with all metrics programs, you must always verify the integrity of the numbers and never trust a single report or someone's assumptions. Always document the "owner" of the metric and who oversees integrity management of the metrics.

Scope: The scope of your evaluation must always be based on ALL elements of your company (e.g., ALL suppliers, ALL systems, ALL buildings, ALL legal entities) and ALL security functions. For example, think about all actual and supporting functions of your security program (for the executive protection function for example, include drivers, intel, travel, medical, close protection, crisis management, training, communications, technology support, etc.). Also, make sure you consider the supply chain (Billy Bob's Limo Service, Mary Mayhem's Medical Services) of your security operations, not just your employee-based functions.

One other note for you to consider as a guideline – I put a timeframe such as "in the last 12 months" for several metrics. This is a guide, do not throw away the metric because you all use another timeframe, such as six months or 18 months. The stronger the constraints of the metric, the better.

Let us Get Started

Top 25 Security Program Maturity Metrics

Security Personnel Management

Functional Description: Does my organization have security professionals that need nurturing? (Uh, look in the mirror). This is NOT a security function; it is a security leadership responsibility as it is all about taking care of your people. No People, No Security Functions. If you implement no other set of metrics in this book, implement these.

One Guy's Perspective: This week, I was getting increasingly excited about seeing *the light at the end of the tunnel* for this book, with my amazing editor making progress, to the cover and website updates, but still nervous about so many areas. But I felt the good Lord remind me that I forgot the most important function of a security department, PEOPLE. So, this morning I sent a note to my editor and asked if I could insert a new chapter that I developed a significant burden for. If we don't take care of our security personnel (employees or contractors), then we are not taking care of our organization's security, period, full stop. I define *personnel* as employees and contractors and recognize that some of the metrics may not be able to be implemented on non-employees, but you can do some of them, even if slightly adjusted.

I love Dr. Bob Nelson's resources on employee motivation and recognition:

I will be the first to admit, and I openly admit it in my mentoring sessions, I am not the best supervisor (by the Human Resources (HR) definition) or not the best at motivation or recognition, thus these metrics challenge me personally also. I try to be a servant leader and have committed myself to an extensive study on honor (hopefully another book that will go from the word processor to physical book one day). This area may also need its own headcount with the special recognition skillset just to plan and execute these ideas and Dr. Nelson's ideas.

Unique Terms:

Honor: The private or public recognition of an individual (usually), typically in a formal way such as a ceremony that may include speakers, physical awards, videos, the blessing of a significant gift, color guards, special music, friends, people the honoree has impacted, etc.

Motivation: An effort by an individual, group of individuals, or an organizational culture to provide another individual or group of people, a strong reason to act or accomplish something.

Recognition: An effort by an individual, group of individuals, or an organizational culture to identify and reward in some minor or major way the efforts of others.

Top 25 Security Personnel Management Maturity Metrics

	The Metric	Best Score
1	Percentage of personnel with access to skip-level discussions on a quarterly basis	100%
2	Percentage of personnel given time off to do pro-bono security work in the community	100%
3	Percentage of personnel given time to present what they are working to an All-Hands	100%
4	Percentage of personnel given time to present/teach on a passionate topic to the rest of the security team	100%
5	Percentage of employees evaluated for "Ice Cream" Awards (e.g., $25 Gift card) on a monthly-basis	100%
6	Percentage of poor performers given opportunity to have one-on-one time with a mentor	100%
7	Percentage of personnel where important dates/tenure/non-project milestones are tracked and recognized publicly (birthday, tenure anniversary, time in security)	100%

	The Metric	Best Score
8	Percentage of personnel in law enforcement or Military or First Responder that are honored and blessed each year on the national days of recognition	100%
9	Percentage of personnel that has all their CPEs paid for by corporate budget	100%
10	Percentage of personnel is offered yearly resume/CV/Credential/Resume Bucket List evaluations by trained professionals	100%
11	Percentage of personnel provided the opportunity to "ride-a-long" with their "dream job"	100%
12	Percentage of personnel where their spouse and kids, if applicable and provided a personal, hand-written note from the CSO on THEIR sacrifice towards the organization's security	100%
13	Percentage of personnel given a sabbatical that is a combination of rest, personal strategy, and organizational strategy (not vacation, paid sabbatical)	100%
14	Percentage of personnel provided financial management resources (e.g., Dave Ramsey classes) to take advantage of	100%

	The Metric	Best Score
15	Percentage of personnel provided opportunity to speak at a local, national, international professional meeting	100%
16	Percentage of personnel provided the opportunity to share their personal talent in "Security's Got Talent"	100%
17	Percentage of personnel provided the opportunity to talk about their favorite charity	100%
18	Percentage of personnel provided the opportunity to red team something	100%
19	Percentage of personnel provided the opportunity to take an awareness class on something total "180 degrees" from their primary role (e.g., guard getting to spend a day in a cyber security operations center (SOC), project manager to go to the range, etc.)	100%
20	Percentage of personnel provided the opportunity to be in security awareness videos	100%
21	Percentage of personnel given an allotment of team Challenge Coins to give to deserve security aware individuals they encounter	100%

	The Metric	Best Score
22	Percentage of personnel that are evaluated for "employee of the _____" (e.g., good parking space anyone)	100%
23	Percentage of personnel provided the opportunity to write their own bio on your website used by the rest of the enterprise (not just your team's internal site)	100%
24	Percentage of personnel surveyed and asked about what would motivate them	100%
25	Percentage of personnel who get to design and implement part of your team's next tabletop exercise	100%
	Extra Credit: Percentage of personnel who get to design and implement part of your team's Escape Room exercise	100%

Executive Protection

(Inclusive of Event/Personal/Close Protection)

Functional Description: Does my organization have executive, travelers, employees, or events that should have added protection when away from the security of their main office? This security function is relevant to all types/sizes of organizations and can engage 24x7 or on an as needed basis.

One Guy's Perspective: I have humbly been involved in the executive/personal/close protection and event security discipline for over 18 years, from a minor scale, in the beginning to progressively more significant operational activities over time. My work has ranged from faith-based, non-government organization/non-profit to Fortune 500, small and large events, and high net worth individuals' activities. My exposure and operations have been as a protector, logistics, gear management, and training lead. My detail lead experience is more limited as I tend to augment teams. Part of these roles has also been to observe and assess teams, support teams, train teams, manage client challenges, and develop people, process, and technology solutions.

Unique Terms:

Protector: Employee or non-employee charged with safety and security of client assets, primarily individuals in-scope of the protection detail.

Principal: The individual assigned to a protector or protector team.

Client: The financial, legal, or operational person or organization supplying requirements (and money, usually) to the protector, protector team, and legal entity employer of the protection team.

Event: An operational period of protection of the principal or other client assets by the protection team for a client under the client's requirements.

Top 30 Executive/Event/Personal/Close Protection Program Maturity Metrics

	The Metric	Best Score
1	Percentage of protectors licensed in the jurisdictions needed	100%
2	Percentage of team members with a comprehensive background check, driving record and chemical testing completed/updated in last 24 months	100%
3	Percentage of time primary vehicles were down due to repairs for lack of maintenance	0%
4	Percentage of time you or your team have documented your advances (paper or system)	100%
5	Percentage of events that have after action reports (AAR) produced	100%
6	Percentage of after-action reports ideas that are >30 days old without implementation	0%
7	Percentage of protectors without minimum state, department, employer, client training completed and documented	0%
8	Percentage of engagements with contracts, reviewed by lawyers AND operations (Brown M&Ms?)	100%

	The Metric	Best Score
9	Percentage of contracts that have MUTUAL confidentiality requirements added	100%
10	Percentage of protectors that are fully up to date on their credential/training renewals and new training requirements. (Do you have a progressive training/growth path documented?)	100%
11	Percentage of principals that have real-time online threat monitoring and analysis produced and delivered to detail lead	100%
12	Percentage of time you have >72 hours of notice to complete a quality advance	100%
13	Percentage of utilization of detail leads and protectors during last quarter (balanced usage and training and recovery)	80%
14	Percentage of protectors that has received fatigue relief in last quarter	100%
15	Percentage of protection (sub-) functions that the team has documented, signed, and team trained since beginning of year	100%
16	Percentage of protection functions and areas of operations that has full measures, metrics, KPIs, and CPIs, documented, updated, and reviewed weekly	100%

	The Metric	Best Score
17	Percentage of events, since beginning of the year, which had a full event debrief, with the principal and/or clients	100%
18	Percentage of protectors' technology that has been through a vulnerability and threat assessment by trained cyber/information security experts	100%
19	Percentage of protectors with recent (<24 months), Minnesota Multiphasic Personality Inventory (MMPI)-type evaluation	100%
20	Percentage of incident-ladened events with fully documented incidents and follow up investigations and training. (Have you fully defined "incident"?)	100%
21	Percentage of clients and partners with NPS measured after each event	100%
22	Percentage of gear bags that pass full gear audit (availability, non-expired, and operationally trained), monthly	100%
23	Percentage of automated external defibrillators (AEDs) with up-to-date licenses, pads, and batteries and tested, monthly	100%

	The Metric	Best Score
24	Percentage of radio management team members, licensed and trained, such as Communications Unit Technician (COM-T) or Communications Unit Leader (COM-L)	100%
25	Percentage of events using COMMERCIAL licensed frequencies AUTHORIZED for use by client organization or primary contractor/provider	100%
26	Percentage of training hours that involves actual principal(s)/client(s)	>0%
27	Percentage of tabletop exercises facilitated by team vs. an independent third-party	50%/50%
28	Percentage of team members and principals/clients that take part in yearly third-party operational assessment	100%
29	Percentage of reservists/subcontractors that took part in some of the training hours of primary team	100%
30	Percentage of team with comprehensive weapons (open carry (OC), Taser, Batton, firearm, etc.) and restraint gear (handcuffs) training and certifications updated in last 12 months	100%

Security Organization's Technology Maturity

Functional Description: Does my organization have technology used by the security departments that is unique to the security department? This function deals with all the classic information technology (IT)-type of functions of managing technology, but specifically ***for*** the security teams.

One Guy's Perspective: When I started in security, and still to this day, I was so surprised how much technology that it takes to run (different than what is used by the security team) a security organization. Of course, the more your organization matures, it is likely your technology needs, and thus infrastructure also grows. These amazing professionals help me every day with regular tools and lots of crazy unique requests. Think about the security department websites, databases, analytics tools, desktops, servers, Internet of Things (IoT), applications (SaaS, local, proprietary, COTS), networks, mobile devices, infrastructure, network, cloud-based technology, etc. Whether the people who manage this technology report to security, another organization, or even a supplier, this function must comply with the best policies, methods, procedures, and security levels. This technology needs its security, capacity, resilience, and availability measured. This is not as much about measuring the security of this technology (that topic is in a different chapter) but more about measuring it as a traditional IT department might measure the technology used to support the business. When we say "technology" in these metrics, remember we are talking about the technology that security department uses to run its operations. We are NOT talking about the technology used by the overall enterprise/company to run the business.

Unique Terms:

Infrastructure: From cabling to racks to networks to servers, it is the *backbone* of all the technology your security teams' use.

Software as a Service (SaaS): Software that is usually running in the third-party data center that runs commercial software used by many companies, but has a specific instance used exclusively for your operations. It is usually running in the cloud.

Commercial off-the-Shelf (COTS) Software: Software that is not internally developed and usually not considered open source. This software's primary advantage is the manufacturer being accountable for updates, enhancements, and some parts of the security of the software.

Top 25 Security Organization's Technology Program Maturity Metrics

	The Metric	Best Score
1	Percentage of technology domains that have a defined, integrated architecture and set of published standards	100%
2	Percentage of technology assets that is centrally managed (e.g., Director under CSO/CISO)	100%
3	Percentage of technology assets' logs that are sent automatically to a centralized collection capability	100%
4	Percentage of resources doing proprietary software development who report to the security technology management team	100%
5	Percentage of technology that is non-standard that has a replacement plan within the next 12 months	100%
6	Percentage of technology that is End of Life (EOL) or End of Service (EOS) that has a replacement plan within the next 12 months	100%
7	Percentage of open-source intelligence (OSINT) and company assets tracked on a geographic information system (GIS) technology	100%

	The Metric	Best Score
8	Percentage of communications tools that support secure logging of communications (remaining 10% would be things like very high frequency (VHF) radios)	>90%
9	Percentage of mobile/battery-powered technology that is human verified, monthly, to be operational	100%
10	Percentage of failure alarms of security technology, resolved within seven (7) days	100%
11	Percentage of unique capabilities developed by the team that have a patent applied for	100%
12	Percentage of open-source technology that has had license compliance verification confirmed in last six (6) months	100%
13	Percentage of technology assets using open-source standards where available (e.g., ONVIF)	100%
14	Percentage of technology assets monitored by a health and status tool (e.g., BigBrother, Vunetrix, etc.)	100%
15	Percentage of technology asset management using a configuration management/trouble ticketing system	100%

	The Metric	Best Score
16	Percentage of technology asset management using a change management process and change management system	100%
17	Percentage of proprietary software development that is compliant with secure coding standards	100%
18	Percentage of technology assets with installed, up to date, monitored, anti-malware technology	100%
19	Percentage of technology assets that has been evaluated to decide if the asset is business/operationally "critical"	100%
20	Percentage of "critical" technology that has been through a disaster recovery test in last six (6) months	100%
21	Percentage of critical technology assets under a hardware maintenance program with quick ship options	100%
22	Percentage of technology assets with latest firmware, security, and non-security patches installed	100%
23	Percentage of critical technology assets with cold, warm, hot backup along with regular restore tests	100%

	The Metric	Best Score
24	Percentage of critical mobile technology that has been tested for physical integrity	100%
25	Percentage of technology support personnel that the organization has trained, and assets evaluated for export control compliance	100%
	Extra Credit: Percentage of critical technology assets with manual alternatives in place, provisioned, and tested (e.g., stack of legal pads anyone)	100%

Security Organizations' Cyber Security

Functional Description: Do we have a concern that our security function's technologies can have cyber security vulnerabilities and attacks (insider threat anyone?)? This function is about making sure you "practice what you preach" and securing your security function's technology from the same threats you are supporting for the overall business. Are those cameras secure? Those nice push-to-talk (PTT) radios patched?

One Guy's Perspective: From the Jerusalem virus response in circa 1990 to successfully bricking (which was not my goal) a significant manufacturer's security camera, through a zero-day exploit I built, to EOL/EOS analytics, security functions/departments' cyber security posture is not something that happens automatically. The security of the technology used by the security department and related security functions needs to be high on the priority list. As a first responder, my instructors and team members often remind me not to become another victim, so I need to take care of myself first. Same with an organization's security, outside of life safety, security of the security function is not something you want to be a victim of. When I have done vulnerability assessments/pen testing (physical and electronic), it is incredible how exploitation of security department technology sometimes is the easiest "in." It is a little bit of "practice what we preach" … policy says 10-character passwords, but your security door only requires a 4-digit pin. This also needs to consider where you outsource technical support to IT or a supplier.

Unique Terms:

Technology: Don't limit yourself to stuff with a keyboard. What about cameras, radios, access control panels, or cars, for example. Don't limit your coverage here, think about "things with wires." When we say "technology" in these metrics, remember we are talking about the technology that a security department uses to run its operations. We are NOT talking about the technology used by the overall enterprise/company to run the business.

Top 25 Security Organizations' Cyber Security Program Maturity Metrics

	The Metric	Best Score
1	Percentage of technology assets with latest firmware	100%
2	Percentage of technology assets with latest software (OA, drivers, middleware, OS, AV, DLP, etc.)	100%
3	Percentage of technology assets under a business continuity/disaster recovery plan	100%
4	Percentage of applications audited monthly to confirm principal of least privilege controls in place	100%
5	Percentage of databases audited monthly to confirm principal of least privilege controls in place	100%
6	Percentage of shared drives audited monthly to confirm principal of least privilege controls in place	100%
7	Percentage of shared accounts with secondary authorization	100%
8	Percentage of applications with technology and machine-based (security) log monitoring	100%

	The Metric	Best Score
9	Percentage of proprietary hardware-based assets with technology and machine-based (security) log monitoring	100%
10	Percentage of technology assets that are EOL/EOS	0%
11	Percentage of technology assets with assigned trustee/custodian	100%
12	Percentage of technology assets with assigned information classification	100%
13	Percentage of technology assets that have been through an independent audit within the last six months	100%
14	Percentage of applications/databases with sensitive records that alert CSO/CISO or their designate, of access to those specific records in near-real time	100%
15	Percentage of third-parties managing assets (regardless of ownership) that have comprehensive and audited (by you) security requirements in their contract	100%
16	Percentage of applications supporting the security organization that have been through a comprehensive application/code reviews on a regular basis and at major changes	100%

	The Metric	Best Score
17	Percentage of cyber security professionals on team that are certified by industry leading associations (International Information System Security Certification Consortium (ISC2), SANS, ISACA, etc.)	100%
18	Percentage of technology assets in a centralized inventory	100%
19	Percentage of networks servicing the security organization's technology assets that has PLP network traffic controls	100%
20	Percentage of technology assets used for undercover/intel gathering security work that has been thoroughly analyzed to verify the unique requirements for this sub-function is implemented, logged, and audited.	100%
21	Percentage of technology assets that have been thorough a tabletop exercise (security AND business continuity) within last six months	100%
22	Percentage of security organization cell phones using a device management technology	100%
23	Percentage of technology and incident response assets geographically mapped for incident management purposes.	100%

	The Metric	Best Score
24	Percentage of team individuals, functions, and unique attributes that are part of an open-source intelligence (OSINT) capability and reporting (threat intel on yourself, your team, your assets)	100%
25	Percentage of technology assets that can generate IOCs where the IOCs are shared intra-company and with industry/government IOC sharing partners	100%
	Extra Credit: Percentage of security functions in organization (assuming de-centralized) that report the above metrics to the CSO/CISO monthly	100%

Loss Prevention

Functional Description: Do we have a retail or warehouse or other environments where theft, breakage, waste of sellable products or supplies can impart of costs and ability to service our customers? Usually thought of in a retail or warehouse setting but no need to limit your review of these metrics if you don't specifically have a retail or warehouse presence. Loss prevention is any actions taken to reduce the amount of theft, breakage, or waste in a business.

One Guy's Perspective: Loss Prevention is not my strongest area of security expertise, but I have rubbed shoulders with experts with extensive experience in some Fortune 50 retailers and had light exposure in several sub-functions. But even learned a lot in this area when helping my son with his (insert any large eCommerce site) businesses and warehouse. If you just take loss prevention and focus on your organization's retail and warehouse/storage areas, these metrics can help. Does your company have a retail store or cafeteria for your employees? Does your church have a book/product store in the church building? Can loss prevention (LP) controls/metrics help your online sales? Absolutely.

Unique Terms:

Shrinkage: A term used to measure the in the earnings of a business due to waste or theft.

Top 25 Loss Prevention Program Maturity Metrics

	The Metric	Best Score
1	Percentage of loss prevention professionals trained and certified in the discipline	100%
2	Percentage of retail/warehouse managers trained in basic loss prevention techniques and reporting	100%
3	Percentage of retail/warehouse employees trained in basic loss prevention techniques and reporting	100%
4	Percentage of "Big Box" retailers in a geographic distance (e.g., 1 mile/1.5 km) from your retail locations that you have engaged with on formal information sharing	100%
5	Percentage of neighbor strip mall stores/neighboring pads of your retail locations that you have engaged with on formal information sharing	100%
6	Percentage of law enforcement agencies servicing your retail/warehouse locations that you have engaged with on formal information sharing	100%
7	Percentage of major e-commerce/shipping services companies (e.g., eBay, Craigslist, UPS, FedEx, USPS, Amazon) that you have engaged with on formal information sharing	100%

	The Metric	Best Score
8	Percentage of flea markets visited in last 90 days looking for "resellers" that could involve your products/parts/stock (hidden AirTag anyone?). *Ode to my late father doing this for his employer and his love for flea markets*	100%
9	Percentage of pawn shops in a geographic distance (e.g., 5 mile/8 km) from your retail/warehouse locations that you have engaged with on formal information sharing	100%
10	Percentage of senior leaders that signed off on agreement on lowest threshold for investigations of a loss prevention case. (e.g., stick of gum or pack of cigarettes or $100 jeans?)	100%
11	Percentage of cases that involve theft, slip & fall, and returns that are analyzed for interconnections/relationships	100%
12	Percentage of cameras, cash registers, access control, time sheet machines at all stores that are running the latest version and configuration for Network Time Protocol (NTP)	100%
13	Percentage of store/warehouse parking lots with cameras on outer edges of parking lot including ingress and egress points	100%

	The Metric	Best Score
14	Percentage of store/warehouse ingress and egress points with logged and analyzed License Plate Reader (LPR) cameras	100%
15	Percentage of stores that hold random cash register audits by investigators not associated with the store.	100%
16	Percentage of stores/warehouses with product detection capabilities that employees and contractors must go through during each egress	100%
17	Percentage of store weight scales that are calibrated, each month	100%
18	Percentage of organization managed transportation for retail/warehouse, with near real time vehicle tracking	100%
19	Percentage of organization managed transportation for retail/warehouse, tamper-evident truck seals	100%
20	Percentage of locations with crime statistics tracked, analyzed with trending and reporting (e.g., CapIndex, ADT, CityProtect, police blotter reports) monthly	100%
21	Percentage of locations geofenced with social media monitoring for robbery/theft alerts	100%

	The Metric	Best Score
22	Percentage of locations with product theft trackers	100%
23	Percentage of locations' loss prevention specialists acting as secret shoppers (testing employees AND detection of crime)	100%
24	Percentage of locations with guard retainers in place	100%
25	Percentage of locations with panic buttons tied to a central station	100%
	Extra Credit: Percentage of locations with specific testing for insider threat detection	100%

Security Assessment & Audit

Functional Description: Do we have concerns about the quality of our security controls? This function is the testing, analysis and reporting of security controls (people, process, and technology and their inter-dependencies) to provide *a moment of time* perspective of the tester or testing team of a specific scope of an organization's assets. I tend to find *assessments* have more operational reporting requirements whereas *audits* tend to have more executive, BoD-level reporting. Key is the assessor's needs to *always* be qualified as independent.

One Guy's Perspective: Soon after I started my security career, I was exposed to security assessments and audits but in a less-than-formal implementation model. Then in my first *head of security* role, I received a call from the interim head of security saying,

> *"Jim, we have four (4) SAS-70 Type 2 audits from the Big 5 (back when there were still five), and the first meeting is on your first day when you come on board."*

SAS-who? 70-What? Is Type 1 easier, or is Type 3 easier, and do I get a vote … Nope. Six (6) months plus one (1) day, now THAT was a formal assessment/audit process.

I learned a ton – what to do and what not to do. A few years later, I traveled to Caracas, Venezuela, with my good friend Israel Vila to do our first international organization security maturity assessment. Again, I learned a lot of Spanish and culture. The top few things I learned from doing comprehensive assessments across most security functions over the last 25 years are:

- Make sure the audience is ready for the results.
- Ensure the scope is clearly defined qualitatively and quantitatively, and there is agreement.
- Make sure everyone understands that this is about building relationships and understanding maturity management, NOT JUST fixing a few findings.
- It is not about the number of findings, but it is about the quality of findings, AND included solutions, are key
- The minute you say, "we are done," it is an out-of-date assessment.
- Compliant does not necessarily mean secure.
- The skills of the assessment team must match the scope
- Security assessments usually have a different audience than audits (e.g., BoD), but they also could be interchangeable
- Security assessments scope needs to cover threats, vulnerabilities, mitigations, and the supporting people, processes, and technology
- Never do the assessment alone, either partner with a peer professional or partner with a trainee or both. Two set of eyes and ears are better than one.

Word of caution on supply chain security assessments

- Supply Chain Security assessments requires additional skillsets over regular security assessments
- Generally, findings must line up with the contract (no contract … different problem) unless completely egregious
- Resistance from suppliers or "send us a checklist" tend to find the suppliers are:
    - Security Stupid
    - Big and Good
    - Scared (non-compliant, cost/pricing mistake)
    - Subcontracting
    - Bluffing
    - Hiding Something

When I use the term *security assessments* in the metrics, I am including both security assessments, security audits, and other types of audits that include security as a subset, for that matter.

Unique Terms:

Independent Assessor/Auditor: The individual trained assessors/auditors were not involved in the design or operations of the target scope of the assessment/audit.

Certified Assessor/Auditor: The individual assessors/auditors are certified by a third-party industry organization for assessing/auditing the scope of the assessment/audit. The individuals almost must be able to prove continuing education and requalification on an industry acceptable timeframe.

Penetration Testing: Also known as Pen Testing; tends to be the covert simulation of criminal behavior and exploitation of known and unknown vulnerabilities and weaknesses in people, processes, and/or technology to prove the probability of in scope threats being successful.

Exploitable Vulnerability: A people, process, and/or technology of a single or string of weakness that evidence can be produced (e.g., picture) proving the weakness was or could easily be used to violate policy or commit a crime or cause a life-safety issue.

Fraud Assessment: In context of these metrics, a fraud assessment is simply a security assessment scoped to include or focus on the threat of insider, outsider, or customer fraud.

Top 25 Security Assessment & Audit Program Maturity Metrics

	The Metric	Best Score
1	Percentage of security assessments with measurable defined scope (Not just words and never with "ALL")	100%
2	Percentage of security assessments with clearly documented elements that are OUT of scope	100%
3	Percentage of personnel charged with security assessments that have been trained regularly, in their specific scope and expertise	100%
4	Percentage of security assessment reports that includes a comparable and measurable impact assessment	100%
5	Percentage of security assessment that includes a comprehensive threat (insider, outsider) assessment	100%
6	Percentage of security assessments included in operational dashboards and BoD reporting	100%
7	Percentage of security assessments that include tabletop exercise (TTX) elements	100%

	The Metric	Best Score
8	Percentage of security assessments that include rewards (think Challenge Coin) for good behavior by the assessed non-security people	100%
9	Percentage of security assessments that include non-security experts included for knowledge, perspective, and context	100%
10	Percentage of security assessments that does not use a checklist for anymore more than a back page reference (there, I said it)	100%
11	Percentage of security assessments that include multiple disciplines of operational security (physical, cyber, fraud, investigations, infosec, etc.) – See TOC for other ideas … hehe	100%
12	Percentage of security assessments that include a change record notification when operational (IT/OT) impact may be involved	100%
13	Percentage of security assessments that include law enforcement notifications where physical security scope includes testing of criminal-minded behaviors	100%
14	Percentage of suppliers that have recurring security assessments (and results reporting) included in your contract with them	100%

	The Metric	Best Score
15	Percentage of customer required security assessments that have completed with reporting	100%
16	Percentage of third-party assessors supplying documented evidence of their training, functional and industry experience for the scope of the assessment you hired them for	100%
17	Percentage of assessors (internal or third-party) that have been trained on history, culture, strategy, and the "Why" of your organization and its mission	100%
18	Percentage of security assessments with managed after action/lessons learned collected and implementation (how do we do better assessments in the future?	100%
19	Percentage of security assessments that include evaluation of findings and recommendation that can implemented in out-of-scope environments to proactively improve the enterprise	100%
20	Percentage of security assessments where the assessor checks if nexus parties (industry peers, co-tenants, etc.) can/should/could be included in partial reporting, to increase value to all parties	100%

	The Metric	Best Score
21	Percentage of security assessments that includes documenting out of scope discoveries	100%
22	Percentage of security assessments with all documentation labelled with proper protection terminology (e.g., classification, attorney client privilege, NOFORN, etc.)	100%
23	Percentage of security assessments reports with individual access watermarks (e.g., User ID of viewer/printer in a watermark)	100%
24	Percentage of security assessments reports with content integrity/changes tracked in a log	100%
25	Percentage of security assessments with evidence protected separately from (summary) reports	100%
	Extra Credit: Percentage of security assessments that includes (budget for) food for the burned-out non-security person you just should take out to lunch for them to brain dump the dirty laundry.	100%

Asset Management (for Security Use)

Functional Description: Do you have assets (non-human) in your organization that you feel need protection from insider and outsider threats? Do you feel your existing asset inventory has all the data elements needed for your security teams' role? Asset management is critical for many business reasons, like finance. This needs to include, buildings, critical *stuff* in the buildings, software, hardware, critical information/documents, supplier information, source code, patents, trademarks, customer/employee information, etc. Security has a critical need to meet its mission, to have high quality, comprehensive asset data/information. Security also tends to need asset data/information that is different than other teams. Operational/Security *ownership* is likely the most important *field* in the database/spreadsheet.

One Guy's Perspective: Risk Assessment (there, I said it; it only took me about 77 pages to drum up the courage to use it … long story) 101 class, *back in the days*, taught all of us that Risk Assessments involve ASSETS, threats, vulnerabilities, AND mitigations in the *formula*. These elements of the formula are hard to scale to get anywhere near 100% coverage for the execs, or the scope is clearly defined as significantly smaller (pie chart anyone), but we keep using the phrase. I am okay with that … but, I will focus the next set of metrics on what I think is the hardest part of the security risk formula (minus politics, money, and humans) – asset management. Without high-integrity, quality asset data, figuring out if your full scope of controls, assessments, defensive coverage, etc. is covered, it is nearly impossible or, at a minimum, non-maliciously deceptive to your audience. Security teams need asset information to count them, secure them, verify their integrity, find them, test them, find their *owner*, and measure their security quality.

The bad guy only needs to find one exploitable asset (human, technical, or physical).

Asset data from IT or Finance can have some value to security functions but be careful when considering their inventory as a gold source for high integrity asset data.

One of the other advantages, and the most vital advantage, to good asset (ownership) data, is to be able to (manually or through automation) assign and thus measure the security posture of assets based on an org chart. (That is for another book).

Stop assigning vulnerabilities, threats, mitigations, and risks to department or business unit names, and start assigning them to individuals. Individuals fund and fix security problems, not departments.

Unique Terms:

Assets: From a security perspective – Something of value owned or managed by the organization, which needs to be protected. I know someone might think a record in a database is an asset, and yes, it is, but don't get so granular that you make these metrics impossible to implement in your first implementation.

Asset Classification: The level (usually 3-4 for each part) of confidentiality, integrity, and availability for that asset to drive control requirements. Some organizations will use other terms like sensitivity, criticality, etc.

Top 25 Asset Management (for Security Use) Maturity Metrics

	The Metric	Best Score
1	Percentage of asset names, types, categories that have senior leadership, CSO, CISO agreement on definition and scope and level of material nexus to risk	100%
2	Percentage of risk assessments that includes asset information/data	100%
3	Percentage of assets track in a non-financial asset management system	100%
4	Percentage of assets with documented assigned trustee/custodian	100%
5	Percentage of suppliers that have your assets in their facilities or their technology, that has CSO/CISO approved security requirements documented in their contract and verified by CSO/CISO within the last 12 months	100%
6	Percentage of supplier assets that are in your facilities or your technology that your CSO/CISO verified compliance to all security requirements, in the contract, within the last 12 months	100%

	The Metric	Best Score
7	Percentage of customer that have your assets in their facilities or their technology that has CSO/CISO approved security requirements documented in their contract and verified by CSO/CISO within the last 12 months	100%
8	Percentage of customer assets that are in your facilities or your technology that your CSO/CISO verified compliance to all security requirements, in the contract, within the last 12 months	100%
9	Percentage of consumer assets that are in your facilities or your technology that your CSO/CISO verified compliance to all security requirements, in the contract, within the last 12 months	100%
10	Percentage of asset names, types, categories with defined and audited destruction requirements	100%
11	Percentage of destruction suppliers that have been audited at their facilities by your CSO/CISO (team)	100%
12	Percentage of non-material assets that have been evaluated for inclusion in your vulnerability management program	100%

	The Metric	Best Score
13	Percentage of non-material assets that increase the vulnerability of organization	0%
14	Percentage of assets tracked with financial replacement cost, fiscal impact on business from being unavailable, financial impact from a breach, and financial management/maintenance costs	100%
15	Percentage of assets with managed end of life/service/support dates	100%
16	Percentage of assets with replacement timeframe being longer than business continuity requirements	0%
17	Percentage of assets owned by a supplier (managing your stuff) checked for compliance based on internal vulnerability/pen test timeframes of similar asset types	100%
18	Percentage of assets names, types, categories with defined vulnerability/attacking testing timeframes (e.g., quarterly)	100%
19	Percentage of supplier owned "wrapper" assets considered a layer of protection for your assets, hat have vulnerability/pen tested based on similar type of assets internally.	100%

	The Metric	Best Score
20	Percentage of assets covered by various risk insurance	100%
21	Percentage of assets affecting national critical infrastructure, which are documented and managed to this level of criticality/sensitivity	100%
22	Percentage of assets affecting human life & safety, which are documented and managed to this level of criticality/sensitivity	100%
23	Percentage of operational roles that are tested by simulation of elimination of critical human assets	100%
24	Percentage of assets that holds export control elements that have been verified compliant within last six months	100%
25	Percentage of assets vulnerable to breach that has no mitigation solution	0%

Law Enforcement Relationship Management

Functional Description: Do you need law enforcement or first responders? Let me answer that, YES. Have you considered how much value add you can be to these heroes, or they can be to you? This function is about building strong relationships and supporting law enforcement and first responders, as how they help your business, locations, and operations. It's more than bringing donuts. Train them, get training from them, Train with them.

One Guy's Perspective: Some of my top non-family heroes are law enforcement officers, firefighters, medical first responders, pastors, and teachers. One of my study topics, outside of security, is on honor and honoring these heroes and others. My amazing wife was a former law enforcement officer, and I get to serve and train with firefighters regularly, including getting my wildland firefighter credentials a couple of years ago. I also have the privilege of mentoring law enforcement officers transitioning to private sector security. If your (security) organization has not built ongoing relationships with law enforcement (and firefighters and medical first responders), this put this book down and start; they are WAY more important than all the other metrics, stories, and solutions in this book.

Relationships between your security team, executives, and law enforcement are critical for so many areas including but not limited to retail intelligence, training support, crime assessments, adverse terminations, active threat response, etc.

Start with building deep relationships with law enforcement (firefighters and medical first responders) and then measure the maturity of your relationship-building program with these simple metrics.

Unique Terms:

Law Enforcement: Local (town, city, county), State, Federal should all be in scope of your work in this area.

First Responder: Member of organizations such as Fire Department, Emergency Medica Services (EMS), Search and Rescue teams, or Water Rescue teams.

Top 25 Law Enforcement Relationship Program Maturity Metrics

	The Metric	Best Score
1	Percentage of local, state, federal "chiefs" covering your location(s) your CSO, CISO, AND CEO have met with to build ongoing relationships	100%
2	Percentage of building/personnel threat tabletops that include local, state, federal law enforcement	100%
3	Percentage of public information officer(s) (PIO)/Media Relations personnel trained and partnered with law enforcement public affairs/PIO peers, particular around joint press questioning	100%
4	Percentage of subpoena compliance personnel trained in affidavit writing and supplying live testimony	100%
5	Percentage of information/data request types include in your law enforcement request guide	>90%
6	Percentage of subpoenas tracked in a secured application/database	100%
7	Percentage of subpoenas included in analytics to discover internal process improvements	100%

	The Metric	Best Score
8	Percentage of company locations where the receptionist/guard stations has updated and tested procedures/post orders to handle a visitor that claims to be a regulator, law enforcement officer, media	100%
9	Percentage of subpoenas included in analytics to discover education opportunities for law enforcement	100%
10	Percentage of buildings threat exercises that involve law enforcement (and first responders)	100%
11	Percentage of high risk (e.g., RF, Electric, Hazmat, etc.) locations where floor plans and tours are attended by law enforcement (and first responders)	100%
12	Percentage of personnel threat exercises that involve law enforcement (and first responders)	100%
13	Percentage of volunteer first responders found within the organization	100%
14	Percentage of systems/applications (and their changes), evaluated for impacts on any information sharing processes (e.g., subpoenas) supporting law enforcement agencies, by trained and certified law enforcement support professionals	100%

	The Metric	Best Score
15	Percentage of supplier contracts that have proper language for immediate notification of law enforcement engagement that impacts your assets	100%
16	Percentage of systems/applications (and their changes) evaluated for impacts on any law enforcement agency's operations by trained and certified law enforcement support professionals	100%
17	Percentage of organization's legal team that have been part of knowledge exchange with law enforcement agencies that affect your organization	100%
18	Percentage of personnel, supporting customers that have specific law enforcement related requirements, that have been trained on these requirements.	100%
19	Percentage of physical pen tests that notifies law enforcement for awareness of false positive 9-1-1 calls	100%

	The Metric	Best Score
20	Percentage of systems, processes, facilities that support customers that have specific law enforcement related requirements, which have been assessed/audited by certified and trained assessment professionals with specialized law enforcement expertise.	100%
21	Percentage of locations with established "retainer" relationships and procedures for hiring off duty (OD) law enforcement officers and other first responder roles, for event support/threat control	100%
22	Percentage of mergers and acquisitions (M&As)/Divestitures/Strategic Projects that include evaluation of impacts on law enforcement (buy or sell-side)	100%
23	Percentage of ethics hotline reports that have been evaluated for possible immediate law enforcement engagement (including at 3am on a Sunday)	100%
24	Percentage of security and safety headcount recruiting that includes reach outs to local law enforcement transition support resources	100%

	The Metric	Best Score
25	Percentage of buildings with response kits/sashes/keys and maps for when law enforcement (and first responders) must respond to an incident (e.g., does everyone agree on what is considered the "North Entrance" and where is the all-access key ... hopefully not with the facility person that is working from home today)	100%
	Extra Credit: Percentage of law enforcement (and first responders) locations you have blessed and honored with breakfast, lunch, or dinner in the last 12 months ... hey and throw in some organizational swag while you are at it!!	100%

Insider Threat Management

Functional Description: Does our organization have insiders? Could they do bad? This function is about the prevention, detection, response, and awareness training for trying to get insiders (employees and non-employees) to not "go down that slippery slope" of unethical behavior, violating company rules, or even criminal behavior.

One Guy's Perspective: I have been involved in insider threat management (formally) for over 20 years. When I started, I and the amazing team, did like most SOC-building teams, we followed the classic IP/*Cyber*-SOC model. Now that I have led the building, operating, and teaching on insider threats, I would do it very differently. These maturity metrics are one of the outcomes of 20 years of working on this problem. I am NOT in denial that SCALING the CONTROLS and post-incident project management for insider threat is the most challenging part of an insider threat program. In 2022, I am still asked about the who, what, when, where, and how of building an insider threat program from private sector to government agencies. It does take some careful governance model and charter agreement before starting an insider threat program, and it must be 24x7.

Unique Terms:

Insider: Individual (employee or non-employee) that has authorized physical or electronic access to company assets that can breach the physical, confidentiality, integrity, or availability of an asset. This typically does NOT include customers.

Top 25 Insider Threat Program Maturity Metrics

	The Metric	Best Score
1	Percentage of CEO and their direct reports that agree upon the scope and ownership of insider threat within the organization, in written form (e.g., Charter)	100%
2	Percentage of insider threat program personnel trained twice a year on internal and specific legal elements of their domain (e.g., ethics, information security, liability, privacy)	100%
3	Percentage of PIO/Media Relations personnel trained in the unique aspects of insider threat briefings	100%
4	Percentage of CEO, CEO Direct Reports, and BoD Members that have read and signed off on the policy/charter granting the right department to be the only department to perform insider threat management/investigations.	100%
5	Percentage of servers, laptops, desktops, mobile devices covered by data loss detection and prevention technology	100%

	The Metric	Best Score
6	Percentage of insider threat personnel (analysts to investigators) that have been through employee assistance program (EAP) training and psychological evaluation and warned of impacts	100%
7	Percentage of investigations that have evaluation criteria and evidence collection procedures to include or rule out insider threat	100%
8	Percentage of company locations where the receptionist/guard stations have been updated and tested procedures/post orders to handle a visitor that claims to be a regulator, law enforcement officer, lawyer, media	100%
9	Percentage of insider threat event data that does not include individual personnel identifier	0%
10	Percentage of personnel that go through annual insider threat prevention, detection, and reporting training	100%
11	Percentage of insider threat personnel (analysts to investigators) that have been through specialized insider threat data analysis training (what if there is no IP address...)	100%

	The Metric	Best Score
12	Percentage of insider threat protection personnel certified in at least one industry recognized investigations certifications (ASIS PCI, CFE, CFI, etc.)	100%
13	Percentage of cyber security pen tests that includes attempts to breach an organizations' assets based on insider threat knowledge	100%
14	Percentage of physical security pen tests that includes attempts to breach an organizations' assets based on insider threat knowledge	100%
15	Percentage of supplier contracts that have proper language for mutual investigations support due to insider threats	100%
16	Percentage of systems/applications (and their changes) evaluated for insider threat vulnerabilities (theft of information/data, fraud, theft, or denial of service(s), etc.) before implementation by trained and certified insider threat professionals	100%
17	Percentage of audits that involves insider threat vulnerability checks (this should be easiest one ever)	100%

	The Metric	Best Score
18	Percentage of personnel, supporting a specific customer, that is trained in the specific insider threat requirements included by the customer in their contracts	100%
19	Percentage of supplier personnel trained in your specific insider threat requirements you included in your contract with the supplier	100%
20	Percentage of insider threat cases/data analysis that triggers after action/control improvement projects	100%
21	Percentage of egress network access points are covered by insider threat event collection systems	100%
22	Percentage of M&As/Divestitures/Strategic Projects that have higher level of insider threat coverage	100%
23	Percentage of ethics hotline reports with an insider nexus that are evaluated for insider threat mitigation	100%
24	Percentage of Bluetooth enabled devices with file transfer capability monitored or disabled	100%

	The Metric	Best Score
25	Percentage of share drives that has had their access control list reviewed and signed off by director or higher in the last three months	100%
	Extra Credit: Percentage of insider threat cases with quoted code of conduct, law or regulatory violations documented	100%

Information Security Management

Functional Description: Does our organization have non-public information? Could it be used for bad consequences against your organization or others? This function is about the prevention, detection, response, and awareness training to protect non-public information, regardless of form or storage mechanism. This is different than cyber security. A manufacturing facility or warehouse with the other "cyber" thing being a time clock still have a lot of information security controls that need to be in place.

One Guy's Perspective: I have been involved in information security management for over 28 years. When I started, it seemed "easy," and today, I am still amazed at how many people, suppliers, and manufacturers make it so difficult, which causes disconnects with the business. Sadly, the industry induced confusion on information security vs. cyber security, doesn't help. I am not denying that SCALING the CONTROLS for information security is difficult … Key management for file cabinets, anyone? Looking back to when I went through my Certified Information Systems Security Professional (CISSP) Boot Camp, back when ISC2 did it over ten days (and something like nine hundred slides), it was so interesting that the elements of a converged security model were embedded within the domains.

Most metrics programs in the information security side of an organization's security management program focus on privacy breaches, basic training, and some basic classification statistics. As with my other *Top 25*, statistics-based metrics programs are still critical, but let us look at these *Top 25* metrics to view the program's maturity, not just some point on the incident, request, or ticket load roller coaster graph.

Though I will make one comment … when it comes to information security, is it an operational function or function that every employee *owns*? Does there need to be a department or person to define the requirements, standards, and guidelines? Absolutely, but should that be a collection of people vs. a single department? Operational and technical ownership of information security will never be centralized, even with someone with the title of CISO. Should SOME of the technical control management be centralized? Absolutely, but that is different from ALL elements of information security management.

Top 25 Information Security Management Maturity Metrics

	The Metric	Best Score
1	Percentage of CEO and their direct reports that agree upon the scope and ownership of information security within the organization, in written form (e.g., Charter)	100%
2	Percentage of information security management personnel trained twice a year on internal and specific legal elements of their domain (e.g., ethics, information security, liability, privacy)	100%
3	Percentage of information assets that are classified but technically cannot met the security standards for the classification	0%
4	Percentage of CEO, CEO Direct Reports, and BoD Members that have read and signed off on the policy/charter granting the right department to be the only department to perform information security investigations	100%
5	Percentage of servers, laptops, desktops, mobile devices covered by information/data loss detection and prevention technology	100%
6	Percentage of storage, emails, and file transfers involving the highest classification of information that has automatic encryption (at rest and in transit)	100%

	The Metric	Best Score
7	Percentage of information assets classified under all three classifications elements (Confidentiality, Integrity, AND Availability)	100%
8	Percentage of file cabinets that have had its protection (lock, fire, B&E) verified and confirmed based on the classification of the information contained within the file cabinets, in the last 12 months	100%
9	Percentage of dashboards that has automated and human checks verifying all dashboard elements/visualizations and exports have principle of least privilege prevention, detection, and corrective controls	100%
10	Percentage of personnel that go through annual information security training that includes hearing and seeing controls (different then cyber security)	100%
11	Percentage of information systems/applications that are covered by a data integrity monitoring tool	100%
12	Percentage of information security investigations personnel certified in at least one industry recognized investigations certifications (ASIS PCI, CFE, CFI, etc.)	100%

	The Metric	Best Score
13	Percentage of buildings secured based on the amount, criticality, and sensitivity of non-public physical information stored in the building	100%
14	Percentage of physical security pen tests that includes attempts to compromise information in the building	100%
15	Percentage of supplier contracts that have proper language for mutual non-disclosure requirements	100%
16	Percentage of systems/applications (and their changes) evaluated for information security vulnerabilities before implementation by trained and certified information security professionals	100%
17	Percentage of audits that involves information security checks (vs. just cyber)	100%
18	Percentage of personnel, that is supporting a specific customer, is trained in the specific information security requirements included by the customer in their contracts	100%
19	Percentage of supplier personnel trained in your specific information security requirements you included in your contract with the supplier	100%

	The Metric	Best Score
20	Percentage of physical investigative evidence having their confidential, integrity, and availability requirements verified based on the right level of classification	100%
21	Percentage of non-public information with authenticated watermark when printed	100%
22	Percentage of travelers given information security awareness before travelling including privacy screens, surrounding awareness, hotel safes, counter espionage, etc.	100%
23	Percentage of internal events with information security announcements/reminders throughout event, including but not limited to, surroundings awareness, public/non-NDA'ed individuals, etc.	100%
24	Percentage of exfiltration control points (cloud, email, proxy, firewall, Bluetooth, etc.) investigated for large and unusual information transfers	100%

	The Metric	Best Score
25	Percentage of public social media posts/blogs, by employees, with a nexus to the organization, which has approval from public affairs, branding, legal approval, and disclaimers	100%
	Extra Credit: Percentage of weight of what the organization sends to the shredder (service) compared to how much paper you	99.999%

M&A Security Management

Functional Description: Do we want to expand throug acquisitions? Do we want to get some synergies or adjus strategy through sales/divestitures? M&A Security is abou taking care of the security and fraud of a transaction from ideation phase through full integration. What security you are inheriting and how do you get it up to your standards Or what security are you delivering to a buyer that won't be embarrassing.

One Guy's Perspective: I have been involved in M&As for about ten years, getting an infamous "Seat at the Table" for security on countless transactions covering billions of dollars. M&A transactions can have the best ROI for a security function but require a unique set of resources and assets to pull it off. I have made a few mistakes in this area and a ton of successes. I know, out of all my *Top 25* lists that I have done and plan to do, this has a much more limited audience, but it works for big stuff like two Fortune ### companies merging as much as it can provide value to two churches joining together, to better serve the community. M&A Security and getting and keeping a "seat at the table" is expensive but so important. What if you help your organization avoid a disaster by helping them with deep-dive security due diligence and discover massive breach or fraud? What if you help your organization reduce the technology integration by migrating security controls faster than building from scratch?

Useful Terms:

Mergers & Acquisitions (M&A): The bringing together of two organizations, in whole or a specific product/service/set of legal entities, through a legal and (usually) financial transaction to form a new company or integrate the one organization into another organization. Yes, legally, and financially, Mergers and Acquisitions can be different, but the operational security elements I have found, are not vastly different.

Sale, Divestiture: The legal, financial, and operational movement of some or the entirety of an organization to another organization; usually the buyer is more prominent in size and scale. Divestitures are sales typically driven by a regulatory challenge or requirement affecting a larger strategy.

Rebadge: The sale or divestiture of part of an organization to another organization, where the receiving/buying organization will continue to perform the same work for the selling party under a Customer/Supplier relationship with the same people (former seller's employees). The term rebadge comes about when the person's badge changes from an employee badge to a non-employee, contractor, supplier/vendor badge. Some have considered this just "simple" outsourcing, but after 20+ years in supply chain security, it is quite different from traditional outsourcing.

… All together now … throughout this chapter when use the term *M&A*, I am talking about transactions that could b an M&A, Sale, Divestiture, or a Rebadge.

Due Diligence: The collection, review, and disposition of information on a company (which could be an individual sole proprietor) that will have access to an organization's non-public assets under a role as a customer, partner, suitor for purchase or being purchased, supplier/vendor/contractor/3rd party/etc. This function is performed to evaluate a target company and its employees' (usually BoD, senior leadership, and investors) trust, integrity, honesty, propensity to commit a crime, fraud, or violation of company policies. Due diligence often has a different business context for people in finance or business development or legal or M&A, so I am going to focus on the security elements which can be independent or part of/nexus to these other teams.

Top 25 M&A Security Management Maturity Metrics

	The Metric	Best Score
1	Percentage of M&A transactions that engage your security team at Ideation Phase (or before)	100%
2	Percentage of M&A transactions where all document storage is secure to both parties' highest standards	100%
3	Percentage of M&A personnel that are "read-in" go through unique confidentiality training	100%
4	Percentage of M&A documentation that has DLP tracking for all documents	100%
5	Percentage of advisor and "sub-advisors" that are under a signed contract from/through BOTH parties to a transaction	100%
6	Percentage of M&A transactions where the security team is engaged for deal-related terminations	100%
7	Percentage of M&A due diligence on the other party (buyer or seller) that includes but not limited to sanctions, PEP checks, Anti-bribery & corruption, financials, reputation, legal issues, fraud issues	100%

	The Metric	Best Score
8	Percentage of documentation and task management is secured by a M&A transaction management tool with strong, certified, and audited security controls (e.g., Midaxo) (hint: Avoid a "shared drive")	100%
9	Percentage of M&A due diligence that includes reviewing quality of employees and non-employees background checks	100%
10	Percentage of M&A Transactions where the data cleanse is completed within 30 days after closing.	100%
11	Percentage of M&A due diligence that also includes due diligence on suppliers and customers being integrated (e.g., KYC)	100%
12	Percentage of M&A due diligence completed on seller's acquired companies/brands	100%
13	Percentage of M&A transactions where both security teams meet regularly (weekly) for cross workstream project management	100%
14	Percentage of transaction's in-scope (critical) buildings reviewed for security vulnerabilities, by the receiving team's security department, prior to deal closing	100%

	The Metric	Best Score
15	Percentage of transaction's in-scope buildings with a completed vulnerability assessment completed within one year before closing	100%
16	Percentage of transaction's in-scope buildings with a completed threat assessment completed within one year before closing	100%
17	Percentage of transaction's in-scope technology elements with a completed vulnerability assessment completed within six months before closing	100%
18	Percentage of M&A transactions that supplies sufficient expense and capital funding for project management, integration management, standards gap mitigation, and ongoing maintenance of security elements	100%
19	Percentage of transaction's in-scope systems, applications, databases, and other devices' vulnerabilities that have been mitigated before close (assumes all these assets have been vulnerability tested recently)	100%

	The Metric	Best Score
20	Percentage of transaction's in-scope physical security vulnerabilities that have been mitigated before close (assumes all these assets have been vulnerability tested recently)	100%
21	Percentage of M&A transactions where Code of Conduct or Investigation Policy/Charter has been implemented at Closing ("Close Day + 1 Minute")	100%
22	Percentage of M&A transactions where network connectivity will be built (acquisition) or restricted (Sales/Divestitures/rebadges) based on allow list/PLP at Closing ("Close Day + 1 Minute") (Extra Credit: And a "go-rack" of security detection/discovery tools is deployed, at the same time)	100%
23	Percentage of M&A in-scope suppliers that have been notified at closing, on changes in security requirements and reporting	100%
24	Percentage of M&A in-scope customers that been notified at closing on changes in security requirements and reporting	100%
25	Percentage of M&A in-scope employees that been notified at closing on changes in security requirements and reporting	100%

	The Metric	Best Score
	Extra Credit: Does your organization keep and support a database of all organizational elements of an M&A transaction and perform regular link analysis of that data to discover non-public interconnections or future concerns? (Think 6 Degrees of Kevin Bacon)	

Fraud Management

Function Description: Do we have financial assets or assets with a financial value that could be abused to commit fraud against our organization or against another organization/person? Fraud management is critical for all organizations. The financial, morale, and brand impact can be significant. Understanding The Fraud Triangle and how individuals fall into it to take advantage of your organizations assets is critical to your ongoing success.

One Guy's Perspective: I have been involved in Fraud Management for over 28 years. It is one area of security that I fell in love with early on, and owe it to two early career mentors, Tom Montuori and John Lewandowski. I wanted my CFE certification before knowing the true and expansive value of certifications. ACFE, its founders and leaders, and my fellow CFEs also hold another special place in my career and "professional heart." One of my special memories is that at one of their conferences, I wrote the first draft of my first patent. Though I do not work on fraud investigations daily, I have several support opportunities each month in my professional and pro-bono work. I have been blessed to work and lead fraud operations for many diverse business areas and types of fraud; it has been and will always be a blast and learning experience especially in regard to human propensity to commit fraud.

Since there is so much to measure in this area, which most of the time is the most discussed and argued set of metrics in the security world, I continue with my *Top 25* theme to focus on operational improvements and not "jump in with the sharks" of quantifying, reducing, justifying fraud (programs). Here are a few bullets of my perspective of some lessons learned that I believe causes measuring this area to be so difficult and controversial.

- "Don't talk about the F Word" – I had such a fantastic time serving alongside a great Fraud investigator named Cheryl Davis; I learned, cried, and laughed so much working with Cheryl, and learned a whole other level of passion for fraud from her. One of the most profound things we talked about what how much "the business" didn't want to talk about "The F Word," … right, not THAT F-word … but talking about FRAUD. Wow, after all these years, still profound and realistic.
- Harder to avoid than classic investigations – One of the struggles in organizations is prioritizing investigations (stolen coffee mug vs. violence). But with fraud, the business and leaders cannot look the other way or minimize its priority.
- Most of the time, it is a control problem. The other struggle with organizations with fraud risks is the reason for the fraud is usually missing, weak, or ignored controls, and thus when fraud is discovered, the work for the business does not end when the case is

closed. The financial loss is not just the fraud, but the work to tighten controls. "Mo Money Mo Money."

- "Fraud is only a finance department problem and isn't security" – If security is the prevention, detection, and response to a crime or violation of organizational policy, that sure sound like fraud to me also. Is there a difference from a business or operational perspective? Is fraud more financial related than stolen coffee mug cases, of course. Is this confusion caused by a security organization or its leaders fears of also managing fraud? Do we need financial and investigative (and … and … and) skill sets together? Yes, but if this concern is affecting your organization's prioritization of fraud, you might not need a *Top 25* quite yet.
- It takes years to investigate and clean up, and that is not at the "speed of business" – Fraud investigations will never be the "normal" open and shut case as the old' stolen coffee mug case. "Just make it go away." Sorry folks, I never experienced a fraud investigation that did not last months, if not years … if done right. And one case or discovered weak control usually triggers more fraud cases. They never run at the speed of business. Talk to a consumer identity theft victim about how long it takes for them to resolve/recover from fraud fully.

Face it, we, as fraud investigators, are humans affecting other humans' lives and the company's future. Fraud Investigations are stressful, but I will say, it is still serious fun … yes, I said it, fun. Some of my most outstanding technical mentors have been fraud investigators (and some were not technology-type technical).

On the business side, I have leaders say, "I wish we never had fraud." I even had a president of a company, where I was doing an investigation, ask me how he could fix the problems, so they do not happen again; my response that popped quickly into my head was, "Stop Hiring Humans." The truest of answers but a priceless sad look on his face. Talking to an investigations manager once, "I found ### fraud cases I am getting ready to turn over to you." His direct response: "You want to know how many of those I want, ZERO." Thus, if you ever take my Analytics 101 class, you now have a hint about Step 1.

Now throw in the need to quantitatively justify the end-to-end fraud management function or prove its value and scope … Wow, just pile in on Jim, why don't you? It is one of the most critical functions in most security departments and one of the most questioned. Confrontation, Employee Morale, Discipline, Breach, Non-Compliance, Congressional Inquiry, Threat, Vulnerability, Termination, and Adverse Termination are not making the top of the list of favorite board game words at the BoD meeting.

Most metrics programs on the fraud management side of an organization's security management program focus on cases opened/closed, category trends, and financial recoveries. As with my other *Top 25*, statistics-based metrics programs are still critical, but let us look at this *Top 25* to view the program's maturity, not just some point on the caseload roller coaster graph.

Though I will make one comment regarding fraud management, this function should be independent, report to General Counsel, have a skilled attorney available 24x7, and be employed by trained and, preferably, certified fraud management professionals and analysts. Many of the fraud investigations I had to clean up or evaluate investigative functions which exposed vulnerabilities in an organization, have been when a non-security leader (usually finance) ran the fraud management function.

Unique Terms:

The Fraud Triangle: The Fraud Triangle states that individuals are motivated to commit fraud when three elements come together:

1. Perceived pressure
2. Opportunity
3. Rationalization the fraud as not being inconsistent with one's values.

(Thanks to Association of Certified Fraud Examiners (ACFE) for the definition.)

Top 25 Fraud Management Maturity Metrics

Note: I recommend you review my *Top 25* for Investigations as the majority of that also relates to investigations that involve fraud.

	The Metric	Best Score
1	Percentage of CEO and their direct reports that agree upon the scope and ownership of "fraud management" within the organization, in written form (e.g., Charter)	100%
2	Percentage of fraud management personnel trained twice a year on internal and specific legal elements of their domain (e.g., ethics, information security, liability, privacy)	100%
3	Percentage of data with indicators of fraud, is analyzed through link analysis for possible "connected dots" of related cases or trends	100%
4	Percentage of data with indicators of fraud, analyzed by humans, using simple, and specialized software for analysis of trends, common causes, and case nexus.	100%
5	Percentage of fraud management team's work reported on a dashboard, specifically looking for investigator or analyst fatigue	100%

	The Metric	Best Score
6	Percentage of CEO, CEO Direct Reports, and BoD Members that have read and signed off on the policy/charter granting the right department to be the only department to perform fraud investigations.	100%
7	Percentage of fraud management personnel that have been trained in affidavit writing and supplying live testimony	100%
8	Percentage of disciplinary actions that require written sign off by a dedicated disciplinary committee based on a documented discipline policy/standard, regardless of whether the fraudster was an employee, supplier, or customer	100%
9	Percentage of fraud types with defined "potential losses" calculations with time limits	100%
10	Percentage of fraud cases with both actual and potential losses calculated	100%
11	Percentage of fraud management personnel that have specialized training in investigations that can involve regulatory or other government reporting or customer reporting	100%

	The Metric	Best Score
12	Percentage of fraud management personnel certified in at least one industry recognized fraud management/investigations certifications (ACFE CFE, ASIS PCI, CFI, etc.)	100%
13	Percentage of fraud management personnel that go through enhanced background checks, financial background checks, social media checks, and chemical testing, yearly	100%
14	Percentage of CEO, CEO Direct reports and other fraud personnel that have compensatory fraud prevention/reduction performance objectives in their yearly performance reviews	100%
15	Percentage of fraud cases that are reviewed for SOX (or equivalent for your country/organization) reporting	100%
16	Percentage of systems/applications/database (and their changes) evaluated for fraud vulnerabilities before implementation by trained and certified fraud professionals	100%
17	Percentage of audits where fraud elements are clearly in the audit's scope and testing done	100%

23	The Metric	Best Score
18	Percentage of financial processes that involve employees, suppliers or customers that are evaluated for fraud vulnerabilities before implementation, by trained and certified fraud professionals	100%
19	Percentage of employee healthcare transactions that go through fraud analytics (even if through a third-party) with near real time dashboard provided to fraud management team	100%
20	Percentage of audit/fraud assessments that consider "salami attack"	100%
21	Percentage of banking transactions that require two human (employee) authorizations	100%
22	Percentage of personnel (employees/contractors) involved in the organization's financial transaction that are required to take 3-5 days of vacation during critical control points (e.g., monthly close)	100%
23	Percentage of donations (outbound, and inbound if non-governmental organization (NGO)) that go through fraud analytics with near real time dashboard provided to fraud management team	100%

	The Metric	Best Score
24	Percentage of customer signup transactions that go through fraud analytics with near real time dashboard provided to fraud management team	100%
25	Percentage of supplier transactions (PO and T&M) that go through fraud analytics with near real time dashboard provided to fraud management team	100%
	Extra Credit: Does your organization manage a database of all organizational elements of all fraud investigations (augmented with public data), and perform regular link analysis of that data to discover non-public interconnections or future concerns? (Think 6 Degrees of Kevin Bacon)	

Security Investigations Management

Functional Description: Have we ever had an actual violation of policy or suspicion of a crime impacting our organization? Do we have a reporting mechanism? Investigations can range from, "who keeps taking my favorite beef jerky from the company frig to full on BoD level fraud and you need to be prepared. Even if you outsource investigations, it is a critical function ever organization must consider before the ZNN truck or federal law enforcement shows up. This function ranges from reporting to evidence collection to interviews to interrogations to admission seeking interviews to reporting to analytics to after action follow-ups.

One Guy's Perspective: I have been involved in Corporate Security Investigations for over 28 years, some of the best human behavior education anyone could hope for. Some exciting, some anger-inducing, some heart-wrenching, some just sad, all stressful but the best PhD-level set of classes I could have ever dreamed for. Face it, we, as investigators are humans affecting other humans' lives. I have even been investigated (and disciplined) for some poor choices when I was young in the industry. Only by grace am I where I am today. I have even been on the "expert witness" side of cases filed by my employer and filed against my employer. Investigations are stressful, even for a great investigator. Some of my most incredible technical mentors have been investigators (and some were not technology-type technical).

Like on the fraud side, which of course is just a type of investigations, some of the same themes:

On the business side, I have leaders say, "I wish we never had investigations." I even had a president of a company, where I was doing an investigation, ask me how he could fix the problems, so they do not happen again; my response that popped quickly into my head was, "Stop Hiring Humans." The most genuine answer, but sad look on his face.

Now throw in the need to quantitatively justify an investigative function or prove its value and scope. Wow, just pile in on Jim, why don't you? It is one of the most critical functions in most security departments and one of the most questioned. Confrontation, Employee Morale, Discipline, Breach, Non-Compliance, Threat, Congressional Inquiry, Vulnerability, Termination, and Adverse Termination are not making the top of the list of favorite board game words at the BoD meeting.

Most metrics programs on the investigation side of an organization's security management program focus on cases opened/closed, category trends, and financial recoveries. As with my other *Top 25*, statistics-based metrics programs are still critical, but let us look at this *Top 25* to evaluate the maturity of the program not just some point on the caseload roller coaster graph.

Though I will make one comment when it comes to investigations, this function really should be independent, report to General Counsel, have a skilled attorney available 24x7, and be employed by trained and, preferably certified investigators and analysts. Many of the investigations I have had to clean up or evaluate investigative functions that exposed great vulnerabilities in an organization have been when a non-security leader (usually an HR leader/generalist) ran the investigation function.

Top 25 Security Investigations Management Maturity Metrics

	The Metric	Best Score
1	Percentage of CEO and their direct reports that agree upon the scope and ownership of "investigations" within the organization, in written form (e.g., Charter)	100%
2	Percentage of investigations that has evidence preservation and eDiscovery experts involved (from the beginning)	100%
3	Percentage of investigations personnel trained twice a year on internal and specific legal elements of their domain (e.g., ethics, information security, liability, privacy)	100%
4	Percentage of investigations where reports, datasets, evidence, interview notes, etc. is centrally stored in a fully compliant and monitored security application and database	100%
5	Percentage of investigation evidence, is secured in a high security, tightly monitored physical (No bottom of the line file cabinet or some generic cloud based shared drive) security containment environment	100%
6	Percentage of investigation data that is analyzed through link analysis for possible "connected dots" of related cases or trends	100%

	The Metric	Best Score
7	Percentage of investigations data analyzed by humans, using simple, and specialized software for analysis of trends, common causes, and case nexus.	100%
8	Percentage of investigations teamwork reported on a dashboard, specifically looking for investigator or analyst fatigue	100%
9	Percentage of CEO, CEO Direct Reports, and BoD members that have read and signed off on the policy/charter granting the right department to be the only department to perform investigations.	100%
10	Percentage of investigators and analysts that have been trained in a witness stand "tabletop" in the last year	100%
11	Percentage of investigators and analysts that have been trained in affidavit writing and providing live testimony	100%
12	Percentage of individuals involved in conducting and supporting investigations that have been trained on international privacy requirements for ALL countries they connect with during their investigations/support role.	100%

	The Metric	Best Score
13	Percentage of individuals involved in conducting and supporting investigations that have been trained on international culture differences for when conducting investigations related to both ALL countries (location and heritage) they connect with during their investigations/support role.	100%
14	Percentage of investigations where HR or Legal is engaged first, that also engages with the security investigation team from the beginning	100%
15	Percentage of disciplinary actions that require written sign off by a dedicated disciplinary committee based on a documented discipline policy/standard	100%
16	Percentage of allegations, past incidents investigations that flow through a prioritization assessment on whether a "case" should be opened and how it will be prioritized.	100%
17	Percentage of cases where information/data integrity is signed off on at different intervals/phases of the investigations, including case closeout obviously.	100%

	The Metric	Best Score
18	Percentage of cases that involve executive/security personnel or other sensitive individuals' that are limited to a smaller set of people then general cases. PLP for sensitive cases.	100%
19	Percentage of investigators and supporting personnel that have specialized training in investigations of union (USA and non-USA) represented personnel and the elements of the specific contracts	100%
20	Percentage of investigators and supporting personnel that have specialized training in investigations that can involve regulatory or other government reporting or customer reporting	100%
21	Percentage of investigations and after actions that is tracked and reported based on the hierarchy of the organization up through to CEO	100%
22	Percentage of investigations evaluated on a regularly (monthly?) basis looking for common/recurring themes of problems and common solutions	100%

	The Metric	Best Score
23	Percentage of investigation personnel certified in at least one industry recognized investigations certifications (ASIS PCI, CFE, CFI, etc.)	100%
24	Percentage of extremely sensitive (defined?) investigations that engaged an independent third-party investigation firm	100%
25	Percentage of investigations where a POST investigation workplace violence probability assessment is completed before discipline	100%
	Extra Credit: Does your organization manage a database of all organizational elements of all investigations (augmented with public data), and perform regular link analysis of that data to discover non-public interconnections or future concerns? (Think 6 Degrees of Kevin Bacon)	

Background Check/Due Diligence Management

Functional Description: Do we want to employ and deal with individuals and companies with the highest integrity? Do our background checks on people and due diligence checks on companies help us *weed-out* integrity issues even if it could be a *big sale*? Background checks, Due Diligence, and Chemical Testing (classically know as *drug testing*) are all tools to try to verify and predict future integrity challenges. "Are they really who they say they are?"

One Guy's Perspective: I have been involved in human and company background checks and due diligence for about 22 years. As a person that been part of the design of this type of function and done background checks and due diligence checks on people and companies and as an evaluator of these functions at various companies, I think this is a great area in which organizations can easily mature and measure the growth of maturity using these metrics.

Let me also make a few potentially politically sensitive statements:

- There may be legal limitations of performing these based on country and citizenship or even down to the state level

- I will not be discussing what would constitute *bad* or *disqualifying*
- Hiring or not hiring someone based on the results of their background or chemical test must not be taken lightly. Same with due diligence, be incredibly careful about (unconscious) bias when deciding disposition.
- I have had people work for me that at some point in their life, where a foreign government or their government, considered them a felony.
- I have professional peers that have a criminal background
- I have done some messed-up things in my life and the only reason I am not where I deserve is grace and forgiveness.
- … but background checks, due diligence, and chemical tests/drug tests … have their place in society and business to protect individuals and an organization's assets.

I also am very aware and have been on the wrong end of this aspect of metrics which is that the existence and the implementation of these types of metrics can trigger some harsh conversations within an organization and have a negative/high consequence impact on individuals, from engineers to senior leadership. BoD: "Hey CEO/CSO/CISO, what do you mean you we're not at 100% across the enterprise on background checks/due diligence."

Unique Terms:

Background Checks: The collection, review, and disposition of information on an individual that will have access to an organization's non-public assets under a role as an employee or supplier/vendor/contractor/third-party/etc. to evaluate an individual's trust, integrity, honesty, propensity to commit a crime, fraud, negatively impact your brand, or violation of company policies.

One note that affects some jurisdictions, is criminal/civil checks need to look for jailed, convicted, charged, arrested, and adjudicated terms, as some government records will use different terms.

Chemical Testing/Drug Testing: The collection, review, and disposition of urine, blood, or other testing models from an individual who will have access to an organization's non-public assets under a role as an employee or supplier/vendor/contractor/third-party/etc. where the use/existence of authorized or unauthorized chemicals or drugs could negatively impact the safety or security of themselves or other assets of the organization.

Due Diligence: The collection, review, and disposition of information on a company (which could be an individual sole proprietor) that will have access to an organization's non-public assets under a role as a customer, partner, suitor for purchase or being purchased, supplier/vendor/contractor/third-party/etc. to evaluate a company's and its employees (usual

BoD and senior leadership and investors) trust, integrity, honesty, propensity to commit a crime, fraud, negatively impact your brand, or violation of company policies. Due diligence, many times has a different business context for people in finance or business development or legal or M&A, so obviously I am going to focus on the security elements which can be independent or part of/nexus to these other teams.

Business-to-Business (B2B): Includes M&A, sales, customers, partners, suppliers, divestiture suitors, etc.

Know Your Customer (KYC) regulations: This is a term used by several regulators, typically in the financial industry to make sure financial institutions verify their customers are not on various watch/embargo lists. I will not address this specifically, but I suspect some of the content and metrics below can be adapted to KYC.

Top 25 Background Check/Chemical Testing/Due Diligence Security Management Maturity Metrics

	The Metric	Best Score
1	Percentage of employees who have had a comprehensive background check completed before or within 30 days of their employment	100%
2	Percentage of employee who have had a follow-up background check completed in the last three years	100%
3	Percentage of employee criminal background checks that included ALL states of country of work, citizenship and visited	100%
4	Percentage of supplier agreements (contracts, purchase orders, payments) that communicate requirements for background checks for their employees (initial and follow ups)	100%
5	Percentage of supplier agreements (contracts, purchase orders, payments) that communicate requirements for chemical testing for their employees (initial, follow up, and unannounced)	100%
6	Percentage of unique titles/roles that have a published background check set of requirements	100%

	The Metric	Best Score
7	Percentage of unique titles/roles that have a published chemical testing set of requirements	100%
8	Percentage of background check requirements that care for/incorporate citizenship, state, national, global limitations	100%
9	Percentage of chemical testing requirements that care for/incorporate citizenship, state, national, global limitations	100%
10	Percentage of supplier agreements that communicate flow down requirements related to background checks	100%
11	Percentage of supplier agreements that communicate flow down requirements related to chemical testing	100%
12	Percentage of B2B sales that have a comprehensive due diligence completed before critical business activities start	100%
13	Percentage of B2B partnerships that have a comprehensive due diligence completed before critical business activities start	100%

	The Metric	Best Score
14	Percentage of B2B supply chain relationships that have a comprehensive due diligence completed before critical business activities start	100%
15	Percentage of substantial risk B2B transactions that triggers the automatic engagement of third-party due diligence experts	100%
16	Percentage of high-risk employee hiring that triggers the automatic engagement of third-party background check and due diligence experts	100%
17	Percentage of B2B transactions that cause an all-country check during due diligence	100%
18	Percentage of employee background checks that involve an online content/social media evaluation	100%
19	Percentage of B2B due diligence checks that involve an online content/social media evaluation	100%
20	Percentage of B2B due diligence checks that involve independent reference checks	100%
21	Percentage of employee background checks that involve independent reference checks	100%

	The Metric	Best Score
22	Percentage of employee background checks that involve independent verification of relevant education, training, certifications, and licenses.	100%
23	Percentage of B2B due diligence checks that involve independent verification of company relevant certifications and licenses	100%
24	Percentage of B2B due diligence checks that include government sanctions, PEP checks on BOTH the legal entities, senior leadership, and influence/control entities	100%
25	Percentage of background checks and due diligences that include checks for involvement in "blood" supply chains and other ESG issues.	100%
	Extra Credit: Does your organization manage a database of all background checks and due diligence (within the limits of privacy policies and laws) and perform regular link analysis of that data to discover non-public interconnections? (Think 6 Degrees of Kevin Bacon)	

	The Metric	Best Score
	Extra Credit: Organization has publicly published background check/due diligence reporting mechanism for the public to inform about concerns (could be a 24x7 Ethics Line)	

Physical Security Management

Functional Description: Do we have physical assets we need to protect? Do we have physical mechanisms that we may or may not own or manage that are used to protect our assets? This function is generally, the prevention, detection, response to threats, vulnerabilities, and incidents against physical assets and/or the physical mechanisms (e.g., buildings, street furniture, etc.) used to protect the assets of the organization

One Guy's Perspective: I have been involved in physical security for over 30 years, as an assessor, pen tester, designer, and operational support resource in multiple countries. Also, let me make it clear right up front when I talk about "security," I am talking about any type of security control (physical, people, technology, cyber, process, etc.). When I talk about physical security, I am talking about controls that protect physical assets (e.g., buildings, stores, street furniture) and the physical assets of an organization inside these controlling mechanisms. Physical Security is as old as the old temple. Physical security of things like information is obviously much newer, when people and organizations started realizing information was an asset that needed to be protected, just like electronic stuff. When I work with people on explaining my professional "why," I talk about securing a particular service capability, they are generally amazed that I don't limit "secure" to any particular threat or function of security whether physical or cyber or information or fraud or other security functions.

Unique Terms:

Street Furniture: I put this here because most people know of physical assets and security of them are generally defined. Street furniture is those boxes, cabinets, coverings you see on the side of the street that is used to support, security, protect underground infrastructure like wires, pipes, fiber optic cable, etc. It's an old telecommunications term that was taught to me by my late father. I call it out here an example of things that are missed in the physical security realm because it just blends in. That is why so much of security is about walking around.

Top 25 Physical Security Management Maturity Metrics

	The Metric	Best Score
1	Percentage of buildings with crime index, Federal Emergency Management Agency (FEMA), Flood, Weather threat, and other quantitative risk measures, updated in last six (6) months	100%
2	Percentage of buildings fully compliant with customer requirements from impacted customers	100%
3	Percentage of physical security vulnerability assessments that tests not just people, processes, technology elements but then all of them together	100%
4	Percentage of physical security devices managed by organization that are running latest firmware/software and secure configuration	100%
5	Percentage of physical security devices that are non-compliant with applicable laws/regulations (e.g., National Defense Authorization Act (NDAA))	0%
6	Percentage of software driven physical security technology without internet protocol (IP) connectivity that is part of technology-based vulnerability discovery program	100%

	The Metric	Best Score
7	Percentage of onsite vulnerability assessments that has a senior leader in a “ride along” role	100%
8	Percentage of new employees trained on physical security vulnerability reporting	100%
9	Percentage of appropriate suppliers with physical security requirements in the contract	100%
10	Percentage of buildings that have had a comprehensive/converged security vulnerability assessment completed in last 12 months	100%
11	Percentage of doors on buildings with block plates over door lock	100%
12	Percentage of physical security technology with anti-malware software installed	100%
13	Percentage of keyed locks have been rekeyed in last two years (full core or key cut change)	100%
14	Percentage of discovered physical security vulnerabilities that are tracked centrally (Database and Dashboard)	100%

	The Metric	Best Score
15	Percentage of discovered physical security vulnerabilities that mitigation and risk ownership (could be different humans) are tracked based on current HR organizational information	100%
16	Percentage of CEO/BoD meetings that have dashboard readouts of physical security vulnerability management and incident metrics	100%
17	Percentage of organization's suppliers that could have a material impact on the organization, that have had a comprehensive/converged security physical security vulnerability assessment completed within last 180 days (by third-party)	100%
18	Percentage of guarded (proprietary, third-party, property owner) buildings where guards are certified in cardiopulmonary resuscitation (CPR)/AED/First Aid/Stop the Bleed/blood-borne pathogens (BBP) and equipment is easily accessible	100%
19	Percentage of buildings with law enforcement agency and fire department assessments and drills completed in last 12 months	100%

	The Metric	Best Score
20	Percentage of buildings with fire warden/floor searchers onsite 24x7 (20:1 ratio is recommended)	100%
21	Percentage with buildings where the utility ingress/egress points are locked (e.g., utility access holes, transformer switches, etc.) (within AHJ limitations)	100%
22	Percentage of buildings with panic buttons tied to 24x7 monitoring center	100%
23	Percentage of security improvement projects tracked in a project management tool by a (certified) project manager	100%
24	Percentage of physical security incidents with an after-action report for security improvements	100%
25	Percentage of physical security professionals certified in Crime Prevention Through Environmental Design (CPTED), physical security professionals (ASIS PSP), and/or International Foundation for Protection Officers (IFPO)	100%
	Extra Credit: Organization has publicly published physical security vulnerability disclosure policy and public vulnerability reporting mechanism	

Security Vulnerability Management

Functional Description: Love Mikko Hypponen's quote/book title of: "If It's Smart, It's Vulnerable." Obviously in a converged security world that can apply to people, processes, technology, buildings, etc. Security vulnerability manage is about the reporting, discovery, creation, mitigation of vulnerabilities in any element of your organization that your organization needs to be secured.

One Guy's Perspective: I have been involved in security vulnerability management for over 27 years, as a practitioner that has discovered (even some COTS zero days), exploited, and as all of us have likely done, caused our share of people, process, and technology vulnerabilities. Also, let me make it clear right up front when I talk about "security," I am talking about any type of security control (physical, people, technology, cyber, process, etc.). And when I am talking about "vulnerability," I am talking about weaknesses in a security control that would allow a malicious human (or a device) that exploits the weakness, to breach the confidentiality, availability, integrity, safety, or security of an organization's asset.

"Jim's finding and counting isn't the problem," well, partly true. Vulnerability management programs, inside organizations, have two major challenges, asset management (who *owns* the asset for fixing a vulnerability to said asset) and scaling analysis and implementation processes. Scaling standards like Common Vulnerability Scoring System (CVSS) and Stakeholder-specific vulnerability categorization (SSVC) should be part of the organization's vulnerability management system.

I also am very aware and have been on the wrong end of this element, that the existence and the implementation of this set of metrics can trigger some harsh conversations within an organization and have a negative/high consequence impact on individuals, from engineers to senior leadership. BoD: "Hey CEO/CSO/CISO, what do you mean you have 10,000 vulnerabilities that have been around for years, and only 1,200 mitigated?"

Scope Challenges in this area include the scope of your evaluation should always be based on ALL elements of your company (e.g., ALL legal entities, ALL business units, ALL suppliers, ALL systems, ALL buildings) and functions. For example, think about all the actual and supporting functions of your security vulnerability management program (Employees, Customers, Suppliers, Visitors, Executives, High-Risk personnel, etc.). So, you cannot do "ALL" in your first round, start somewhere, just make sure your audience is aware of the limited scope.

Because this is one area of security, we don't want our audience to be deceived, also think about how to communicate the scope and limitation in the adage about Knowns and Unknowns when it comes to your assets (e.g., is Company software on employee's personal phones, tablets or computers, a company asset?)

	Knowns	Unknowns
Knowns	***Known Knowns*** *Things we are aware of and understand.*	***Known Unknowns*** *Things we are aware of but don't understand.*
Unknowns	***Unknown Knowns*** *Things we understand but are not aware of.*	***Unknown Unknowns*** *Things we are neither aware of nor understand.*

Unique Terms:

Vulnerability Discovery: The discovery of weaknesses in buildings, people, processes, or technology that can be exploited to breach an asset. A discovery doesn't necessarily mean that person doing the discovery needed to exploit the vulnerability.

Vulnerability Scanning: Generally, the automated capability to discover vulnerabilities in technology.

Site Survey: A broad combination of compliance, standards, and vulnerability discovery activities at a physical location.

Vulnerability Exploitation Exercise (Pentest or pen test): The discovery and actual exploitation of discovered vulnerabilities (physical, people, process, or technology).

Compliance/Standards Gap Assessment: An assessment looking at an asset to see if the asset is non-compliant to a defined set of requirements (e.g., a contract) or a set of defined (industry or proprietary) standards. This type of assessments does NOT necessarily discover or exploit actual security vulnerabilities.

Top 25 Security Vulnerability Program Maturity Metrics

	The Metric	Best Score
1	Percentage of organization's brands, domains, IP Space, apps with updated communications information available to Carnegie Mellon's Computer Emergency Response Team Coordination Center (CERT/CC) and their international peers	100%
2	Percentage of contracts that are fully compliant (according to the customer) to customer vulnerability management requirements	100%
3	Percentage of customers receiving regular (e.g., Monthly) communications on customer "owned" vulnerabilities	100%
4	Percentage of customer owned/leased electronic devices managed by organization running latest firmware/software and secure configuration	100%
5	Percentage of desktops/servers (non-appliance) with file-level inventory centrally tracked (for EOL/S and patch management needs)	100%
6	Percentage of software driven technology without IP connectivity that is part of vulnerability discovery program	100%

	The Metric	Best Score
7	Percentage of organization's fleet cars with latest firmware	100%
8	Percentage of organization's fleet with remote connectivity disabled	100%
9	Percentage of financial transactions that have at least one automated fraud test before the transaction is approved	100%
10	Percentage of onsite vulnerability assessments that has a senior leader in a "ride along" role	>10%
11	Percentage of vulnerability assessments testing not just people, processes, technology elements but then all together	25% each
12	Percentage of (full-time employees (FTE) and third-party) external communications employees, trained on developing and making security statements related to public vulnerability communications	100%
13	Percentage of personnel that have been tested against email, phone, instant message (IM)/direct message (DM)-based social engineering during regular vulnerability discovery	100%
14	Percentage of new employees trained on vulnerability reporting	100%

	The Metric	Best Score
15	Percentage of suppliers with physical or logical/electronic access to your organization's assets, where all their personnel involved, are trained on vulnerability reporting	100%
16	Percentage of organizational owned/leased buildings that have had a comprehensive/converged security vulnerability assessment completed in last 12 months	100%
17	Percentage of doors on organization owned/leased buildings with block plate over door lock	100%
18	Percentage of physical security technology included in organization's technology vulnerability/mitigation program	100%
19	Percentage of keyed locks have been rekeyed in last two years (full core or key cut change)	100%
20	Percentage of discovered vulnerabilities that are tracked centrally (Database and Dashboard)	100%
21	Percentage of discovered vulnerabilities that mitigation and risk ownership (could be different humans) are tracked based on current HR organizational information	100%

	The Metric	Best Score
22	Percentage of CEO/BoD meetings that have dashboard readouts of vulnerability management metrics	100%
23	Percentage of organization specific code that has gone through static and runtime code vulnerability scans	100%
24	Percentage of organization's suppliers that could have a material impact on the organization, have had a comprehensive/converged security vulnerability assessment completed within last 180 days (by third-party)	100%
25	Percentage of supplier-managed organization owned/leased devices that has measurable and verified vulnerability management requirements in the signed contract	100%
	Extra Credit: Organization has publicly published vulnerability disclosure policy and public vulnerability reporting mechanism	

Organization Security Training Management

Functional Description: Do you think we should education our employees, suppliers, and customers about safety and security? It has been said that if you were given USD$1 for security, you should spend it on security awareness/training of your people. Training comes in so many forms when it comes to security, but what we focused is getting some "muscle-memory" into personnel now, so when the time comes to act, they don't "freeze." Whether a 5-minute video or several-week course. Never Stop Training.

One Guy's Perspective: I have been involved in security training for over 27 years, as a student, speaker, teacher, practitioner, training lead, and evaluator. My work has ranged from Faith Based, NGO to Fortune 500, one-on-one training, university, professional organizations, mentoring, law enforcement, military, USA, and multiple other countries…. Blessed with a wide range of audiences. Measuring "did the training help security" is hard and sometimes does not manifest its success for months if not years. "I was in a class with you a couple of years back and you said to call you _____." Measuring (security) personnel training does not necessarily mean it will mature your security program in your weakest areas. But overall, training moves the needle on overall security improvements. The hard part is the ROI over the long haul. But my peer Donnie Saunders summarized our insider threat work in such a profound statement, "We provide awareness to try and change people's BEHAVIOR." How we do encourage people to ethically stop at no-more-bad than taking an office pencil ("The Ethical Stop") … you train them.

Unique Terms:

License: Based on a demonstration of skills and knowledge a government agency would issue a license grant rights, authority, and responsibilities on an individual in a particular domain.

Certification: Based on a demonstration of skills and knowledge a company or industry association or NGO would issue a certification that communicates the holder's demonstrated working knowledge of a topic.

Top 25 Organization Security Training Maturity Metrics

	The Metric	Best Score
1	Percentage of new employees trained on baseline set of converged security curriculum within 15 days of onboarding	100%
2	Percentage of suppliers with physical or logical/electronic access to your organization's assets, where all their personnel involved, are trained using YOUR physical security requirements, within 15 days of final contract signature and yearly	100%
3	Percentage of suppliers with physical or logical/electronic access to your organization's assets, where all their personnel involved, are trained using YOUR information/cyber/logical security requirements, within 15 days of final contract signature and yearly	100%
4	Percentage of executives trained on information security (public discussions, BEC, social engineering mainly)	100%
5	Percentage of executive administration team trained on information security (public discussions, BEC, social engineering mainly)	100%

	The Metric	Best Score
6	Percentage of executives trained on (physical & online) personal threat detection/reporting	100%
7	Percentage of executives trained on working with executive protection teams	100%
8	Percentage of (employees and third-party) external communications employees, trained on developing and making security statements and incident response and security PIO management	100%
9	Percentage of customers receiving regular (e.g., Monthly) training on customer owned rules (e.g., incident reporting, fraud, software patching, social engineering).	100%
10	Percentage of personnel with physical/electronic access to organization assets that receive social engineering training involving physical, phone, and online attack vectors	100%
11	Percentage of organizational owned/leased buildings where ALL regular building occupants are trained in active threat response (Run, Hide, Fight)	100%
12	Percentage of software developer personnel, developing organization specific code, are trained in secure coding practices	100%

	The Metric	Best Score
13	Percentage of buildings with at least two certified or licensed thus trained medical response personnel onsite when office is open to when a substantial number of personnel are onsite	100%
14	Percentage of buildings where security personnel are trained quarterly in medical response (CPR/AED/FA/BLS/StB/BBP, at a minimum)	100%
15	Percentage of buildings where all security personnel are trained quarterly in medical Hazmat response	100%
16	Percentage of buildings guarded by licensed guards and/or onsite security personnel, where all personnel performing guard services, are certified/trained in incident response (CPR, AED, FA, BLS, StB, BBP, Use of Force, Evac, Active Threat, etc. at a minimum)	100%
17	Percentage of retail locations (company owned, franchise, indirects) where all employees, are trained in incident response (CPR, AED, FA, BLS, StB, Active Threat, etc. at a minimum)	100%

	The Metric	Best Score
18	Percentage of locations where incident response training includes local law enforcement, fire department, and tactical response teams (e.g., SWAT)	100%
19	Percentage of personnel involved in financial transactions are trained on fraud prevention, detection, and reporting	100%
20	Percentage of incident response training that includes actual (third-party) facilitated tabletop elements	>25%
21	Percentage of internal and external auditors trained in basic investigative interviewing techniques	100%
22	Percentage of security personnel trained yearly in basic soft/integrity skills (interviewing, bias, ethics, etc.)	100%
23	Percentage of marketing/executive event management team members trained in security engagement and event security	100%
24	Percentage of security related licenses and certifications tracked in a database for risk management, incident response, and training assets	100%

	The Metric	Best Score
25	Percentage of authorized personnel verifying and training monthly, on personal, event, mobile, static incident response gear kits	100%
	Extra Credit: Percentage of security training that is accredited (certified, CE/CPEs, etc.) and trained by certified/accredited trainers	>50%

Security/Fraud Analytics (to find bad) Management

Functional Description: Do we have information or data that we could use to help us uncover security issues? Do we have the skill sets to exploit this data? Analytics is about analyzing information and data, using humans and machines to answer initial questions or what I call security/fraud theories and typically forces more questions. Analytics can be done to find threat, vulnerabilities, and breaches.

One Guy's Perspective: I have been involved in data/information/security/fraud analytics (the kind that finds harmful stuff) for over 25 years, to where, now, it is part of the first thing I talk to people about when starting a new security or fraud project, function, or department. Since this book is about metrics to improve security and hopefully the analysis on top of the metrics, this chapter is focusing on maturing your security/fraud analytics that you (hopefully) use to find funny, SMH, scary, and even terrible things, people, activities, vulnerabilities, threats, even previous unknown breaches. I have been teaching a class, which I now called Analytics 101, that tried to educate the students/audience on how best to implement analytics within their organization regardless of input/output and tool. Though I love Tableau and i2 Analyst Notebook, Excel and scripts, human analysis will never be taken out of the analytics toolbox. The variety of information and data sets and of course questions and bad, are endless. Since I am more focused on the maturity of your program, I will redirect you to the amazing folks at ACFE for countless examples of data analytics finding bad. In 2024, I may share some "targets" of data and indicators of bad (IOB … hehe) in a separate section.

Unique Terms:

Reporting: The communication of results of gathering information/data, generally, to tell WHAT happen

Analytics: The communication of results of gathering information/data, asking questions of the information/data to tell the story of what and why something happen or will happen.

Visualization: The communications of reporting or analytics in a, typically, graphical way to draw an audience quicker to answers to questions about the information or data being reported or analyzed

Link Analysis: A type of analytic and visualization that "connects the dots," think old time crime show where they have string between evidence, pictures, victims, and suspects. Obviously, doing this with data using link analysis software can expedite answer gathering much quicker.

Top 25 Security/Fraud Analytics (to find bad) Maturity Metrics

	The Metric	Best Score
1	Percentage of senior leaders that have "bought-in" to the potential outcome of the analytics	100%
2	Percentage of data sets that have been evaluated for 100% integrity quality before being used analytics	100%
3	Percentage of analytics output that is secured to PLP, ACP, PII, HIPAA, etc. standards	100%
4	Percentage of insider threat analytics that includes human attribution	100%
5	Percentage of suppliers with access to organization assets that are contributing data to your analytics program	100%
6	Percentage security/fraud analytics professionals fully trained in analytics principals	100%
7	Percentage security/fraud analytics professionals fully trained in the analytics tools.... certification is encouraged	100%
8	Percentage of quantitative risk models that include human/"gut" analytics	100%
9	Percentage of quantitative risk models that support agreed upon weighting	100%

	The Metric	Best Score
10	Percentage of visualizations/reports/dashboards that include a scope (inclusive and exclusive) statement on each screen/page	100%
11	Percentage of visualizations/reports/dashboards that include a data integrity notification statement on each screen/page	100%
12	Percentage of analytics data storage and output that cares for mixed classification of information/data	100%
13	Percentage of analytics where formulas are fully documented	100%
14	Percentage of reporting, analytics, visualizations that have been analyzed and adjusted for color challenged audiences	100%
15	Percentage of reporting, analytics, and visualizations that have been analyzed and adjusted for visually deceptive visualizations (e.g., pie charts, axis ranges, dual axis)	100%
16	Percentage of business units/legal entities covered in the analysis and output	100%
17	Percentage of analytics reporting based on human organization ownership vs. business unit/brands/legal entities	100%

	The Metric	Best Score
18	Percentage of analytics that is formed from a security or fraud question or theory vs. just analyze data aka data driven	100%
19	Percentage of analytics that meets the organization's retention policies	100%
20	Percentage of investigators using analytics, that have been formally and mock tested in testifying about the findings of their analytics work	100%
21	Percentage of analytics professionals, that support investigators, that have been formally and mock tested in testifying about the findings of their analytics work	100%
22	Percentage of fraud analytics output communicated using Dr. Cressey's Fraud Triangle	100%
23	Percentage of organization's analytics professionals trained in security/fraud reporting procedures	100%
24	Percentage of analytics tools that go through False Positive, False Negative and "Deep Fake" data to test the tools and its configuration's quality and integrity	100%

	The Metric	Best Score
25	Percentage of analytics tools that go through False Positive, False Negative and “Deep Fake” data to test the analytics professionals’ quality of detection	100%

Security Budget Management

Functional Description: Do we have a separate and distinct budget for security labor, non-labor, and capital items? Do we have a process to make sure security budget is include in all new projects? This is classic establishment, sustainment, and tracking of expense, capital, labor, non-labor for security improvements and ongoing operations.

One Guy's Perspective: When I started working at my first job, I ended up being charged with accounts receivables for a small retail store. In another job, I had a customer/insurance collections role for a doctor, that's where I learned patience and perseverance … hehe. I started being responsible for a security budget a few years into my security career. It was small, but it was nice having a budget manager from finance helping me out monthly and quarterly on a line-item basis. Whether your organization runs their bookkeeping off of pen and paper or Quickbooks or a massive Enterprise resource planning (ERP), you need to separate you budget for security. Understanding who *holds* the budget vs. who *uses* the budget is also something to keep track of. So, you set the budget for 202x and new business Project A comes up on January 15, make sure all the expenses/capital for Project A are *tagged* with Project A project codes. Finance needs to understand how to *forgive*, *sweep*, adjust current view (CV) based on these *tagged* expenses for Project A-Z.

Common places that budgets for security and fraud have been *housed*, obviously lots more organizations could be uses it:

- Real Estate
- Information Technology
- Legal
- Finance

Unique Terms:

Capital aka CapEx: Capital expenditures, which is money used by an organization to purchase, improve, or sustain physical assets

Expense aka OpEx: Operating expenses or expenditure, refers to the costs incurred by your business via the production of goods and services

Top 25 Security Budget Management Maturity Metrics

	The Metric	Best Score
1	Percentage of organization's security budgets that has been inventoried	100%
2	Percentage of executive protection budget that has been reported in compliance with regulatory requirements (e.g., in USA that would be IRS and SEC)	100%
3	Percentage of security budgets that carve out new training, sustainment, and CPE support	100%
4	Percentage of new product projects that include a request from various security functional owners for their input	100%
5	Percentage of internal projects that include a request from various security functional owners for their input	100%
6	Percentage of M&A projects that include a request from various security functional owners for their input	100%
7	Percentage of large reduction-in-force (RIF) programs that include a request from various security functional owners for their input	100%

	The Metric	Best Score
8	Percentage of large hiring projects that include a request from various security functional owners for their input	100%
9	Percentage of new customer contracts that include a request from various security functional owners for their input	100%
10	Percentage of new supplier contracts that include a request from various security functional owners for their input	100%
11	Percentage of new real estate contracts that include a request from various security functional owners for their input	100%
12	Percentage of security budgets that have "Charge-back" capabilities	100%
13	Percentage of supplier agreements with financial penalties (CPI) for non-compliance security issues, breach, or security vulnerability issues	100%
14	Percentage of (non-security) operational budgets that have line items for security, fraud, insider threat, business continuity response costs	100%
15	Percentage of security budgets that has line-item level reporting available, at least monthly	100%

22	The Metric	Best Score
16	Percentage of security budget(s)/line items that can be provided in a near-real-time dashboard	100%
17	Percentage of investigation "recoveries" that reverts to the security investigation team for improvements ("finder's fee")	>10%
18	Percentage of organization's supporting functions that can be used by security teams without affecting security's budget (within reason). Example: Training department, content and video department, IT (sysadmin, DBA, etc.)	100%
19	Percentage of capital projects that include ongoing security budget for operations and maintenance in the business case	100%
20	Percentage of security budget cuts signed off by General Counsel, Financial Risk Management Officer	100%
21	Percentage of security budgets with a line item for rewarding good security behavior of personnel	100%
22	Percentage of supplier budgeted security requirements that is evaluated by your organization's security teams for opportunity for insourcing/savings	100%

	The Metric	Best Score
23	Percentage of security budgets with a line item for outreach to community, law enforcement, other government agencies, and non-profits	100%
24	Percentage of security budgets with a line item for donations to worthy security related causes/foundations (e.g., ACFE Foundation, ASIS Foundation, Fallen Officers, etc.)	100%
25	Percentage of security budgets with a line item for resources for security team members volunteering their security expertise to schools and non-profits	100%

(Technology) Supply Chain Security Management

Special Notice: In 2021 and 2022, I had the fantastic opportunity to work with a group of supply chain experts on TIA's QuEST Forum and Supply Chain Security (SCS) 9000. My little contribution was around metrics, and one of the resources was the original *Top 25 for Supply Chain Security*. The list below is a 2023 Edition, but this could not have been possible without the SCS 9000 team and volunteers at TIA. Please check out and implement their work on SCS 9000!

Functional Description: Do we buy product and services from third-parties? Do we rely on third-parties for some part of our security, fraud, risk controls? Supply Chain (Security) is such an amazing, vast, and fun topic. I worked with one company about 30 days after they were formed to show the owner that they already had about 30 suppliers in their supply chain that impacted their security, risk, and operations.

One Guy's Perspective: A group of international experts on the information and communications technology (ICT) supply chain, under the auspices of the TIA QuEST Forum, supported me in the development of my first *Top 25 Supply Chain Security Measurements* to challenge its members, partners, the ICT industry, and the supply chain of other critical infrastructures. It was such an honor to support my first industry standard. When I really started dealing with supply chain security about 22 years ago, it was part contracts, part compliance, and part due diligence. Over the years, my brother Bob and my son James, Jr., and many supply chain experts have imparted supply chain wisdom and knowledge to me. I always love the discussions with these professionals, and I am always amazed at what dynamic challenges a supply chain can throw at you. This set of metrics below is covering ONLY supply chain challenges for the technology supply chains. In the 2024 edition, I will get more into security metrics of other supply chain sub-disciplines.

Unique Terms:

Supplier, third-party, vendor: Any external entity that you get products or services from that affect the risk to the company, regardless of whether you pay them directly or indirectly or have them under contract or not.

Personnel: Individuals, whether employees, supplier/third-party/vendor employees.

Software: Firmware, disk-based software, cloud-based software, operating system level, database level, application level, etc.

Subcontractors: Any external entity, which your direct suppliers use, that you get products or services from that affect the risk to the company, regardless of whether you pay them directly or indirectly or have them under contract or not.

Top 25 Supply Chain Security Measurements

	The Metric	Best Score
1	Percentage of suppliers that have contractual required security requirements	100%
2	Percentage of third-party software installed, which is end-of-life	0%
3	Percentage of third-party software installed that has all security patches installed	100%
4	Percentage of personnel (non-employees) with access to our email environment, which passed our phish test	100%
5	Percentage of third-party managed, internet facing devices, monitored 24x7 for security incidents	100%
6	Percentage of proprietary internal software, developed by third-parties, which has been through an automated and human code review	100%
7	Percentage of third-party managed buildings with our assets, which does not have cameras, access control covering our assets.	0%
8	Percentage of our systems/applications that have suppliers managing the security controls	0%

	The Metric	Best Score
9	Percentage of suppliers that have access to our data, which have not been assessed/audited onsite by our auditors and security teams	0%
10	Percentage of supplier employees accessing our network from unauthorized (per contract) locations	0%
11	Percentage of suppliers that use subcontractors with physical or electronic access to our assets	0%
12	Percentage of supplier employees with access to our assets, which have not had a human-verified background check/chemical test (as allowed by law) ... in the last two years	0%
13	Percentage of COTS devices on the internal network, which are managed, inventoried, with a named owner	100%
14	Percentage of suppliers that have been screened against sanctions listings	100%
15	Percentage of third-party managed email domains supporting, our organization controlled, by DomainKeys Identified Mail (DKIM)	100%

	The Metric	Best Score
16	Percentage of third-party managed Domain Name System (DNS) servers supporting our organization with Domain Name System Security Extensions (DNSSEC) implemented	100%
17	Percentage of supplier employees protecting our environment by performing security functions, not defined in their contract	0%
18	Percentage of suppliers that have 24x7 incident response (IR) contacts in our CSO/CISO's IR program database	100%
19	Percentage of suppliers that support our business and government customers, that we have verified onsite, are fully compliant with the flow down security requirements	100%
20	Percentage of third-parties with physical and/or electronic access to our environment, which is covered by our insider threat monitoring capabilities	100%
21	Percentage of supplier hosted data where access reporting timelines are verifiably within regulatory maximums	100%
22	Percentage of third-party providers that manage security (physical or logical) without contractual managed CPIs/KPIs	0%

	The Metric	Best Score
23	Percentage of shipments (inbound, outbound, third-party products, our products) that are not supplying near real time GPS/RTK package tracking	0%
24	Percentage of suppliers without a financial stability report on file with us, within last two years	0%
25	Percentage of suppliers that affect our business continuity, which have not documented and tested their business continuity plan	0%

Security Metrics for Board of Directors

One Guy's Perspective: I have never been in a Fortune ### BoD meeting; I was a member of a great organization called NACD when I was on a small BoD. But I researched several security *tools* for the BoD and found several good educational articles and "Questions I would love to ask my Board" and "Questions I would love my Board to ask me." Honestly, what I read did not seem measurable (minus some statistical roller coaster stats), and they did not seem hard to answer … they did not move the needle of improving security, and they did not scare the BoD.

So, I hope these maturity metrics challenge and scare the BoD, senior leadership, and the CISO/CSO leadership. The BoD needs to ask these (and better) security questions regularly. The security teams need to be prepared with these regardless of whether you are presenting to the BoD.

With recent interest of USA's Congress on holding public committee meetings with CEO, CSO, CISO, and whistleblowers, I think this is an area that I would be quite worried about not having ***great*** and measured evidence readily available if I was a board advisor to a company.

Top 25 Security Metrics for Board of Directors to Evaluate

	The Metric	Best Score
1	Percentage of our organization's technology assets that will be EOL/EOS by end of the fiscal year	0%
2	Percentage of our organization's buildings that have vulnerabilities that cannot be mitigated by end of the fiscal year	0%
3	Percentage of our organization's suppliers that have contractual non-compliance that will not be mitigated by end of the fiscal year	0%
4	Percentage of our business customers' that have security requirements that we have not met and verified by the customer or their designated third-party	0%
5	Percentage of breaches, regardless of size and type, that have been reported to the BoD each quarter	100%
6	Percentage of BoD members that have signed off on the definition and scope of security, vulnerability, breach, fraud, material, reportable, etc.	100%
7	Percentage of fraud losses, regardless of size and type, which have been reported to the BoD each quarter	100%

	The Metric	Best Score
8	Percentage of investigations that have involved director or higher that have been reported to the BoD within 72 hours of the accusation	100%
9	Percentage of M&A, divestiture, sales, rebadges that involve sign off by the CSO and CISO at due diligence, transition planning, Go-No-Go close and full integration	100%
10	Percentage of BoD members' that have a document explaining all the security (physical, personnel, cyber, etc.) protecting the BoD member, that the organization's security team is responsible for	100%
11	Percentage of organization's security operations reporting metrics, KPIs, CPIs, NPS to the CSO and/or CISO daily	100%
12	Percentage of organization's security governed by the CSO and/or CISO	100%
13	Percentage of M&A, divestiture, sales, rebadges that fully funded CSO and CISO requirements	100%

	The Metric	Best Score
14	Percentage of supply chain (prime and subs), regardless of whether there is a contract, has been through a business risk, corruption, and financial due diligence within the last year.	100%
15	Percentage of supply chain investigations, affecting the organization, which have been reported to the CSO and/or CISO	100%
16	Percentage of BoD reportable audits that checked for data integrity issues	100%
17	Percentage of BoD reportable audits that cover both whether an asset is secure, not just compliant	100%
18	Percentage of BoD reportable audits where the scope and testing specifically includes breach and fraud discovery	100%
19	Percentage of audit and investigations after-action findings that have an assigned owner and funding	100%
20	Percentage of BoD, while performing duties for our organization, have personal protection duties that are performed by trained, equipped, and audited personal protection professionals	100%

	The Metric	Best Score
21	Percentage of BoD meetings that include facility "bug sweeps" and results provided to the BoD before the meetings	100%
22	Percentage of Annual meeting and BoD meetings covered by a social media threat monitoring capability before, during and after the meetings with reports going to the BoD members	100%
23	Percentage of breaches, vulnerabilities, threats that have an attribution nexus to known terrorist countries	100%
24	Percentage of BoD members that have taken part in threat exercises/tabletop exercises in last 12 months	100%
25	Percentage of BoD primary employer's security teams that have met with our organization's security team in last 12 months	100%
	Extra Credit: Percentage of meeting resorts/event locations/golf courses/etc., used for BoD meetings/retreats, that have been pre-meeting audited for security vulnerabilities and threats along with BoD briefings	100%

Security Metrics for Congress/Agency Secretary to Ask Government CSO/CISOs

Note: I am using USA-centric government terms (e.g., Congress vs. Parliament), but hear my professional heart, these work in other forms of government around this amazing earth.

One Guy's Perspective: I have never been in a Congressional (or Parliament) office or on a Congressional floor. I got close, and I have watched C-SPAN on several occasions. Outside my wife, son, and I talking about politics inside our home, I hold my political perspectives close. I like watching the headlines and soundbites with a strong awareness of the consistent audience manipulation activities. I read *How to Kill 11 Million People* by Andy Andrews. But two *C-SPAN* ah-ha moments caused me to include this chapter.

1. I watched three Congressional leaders incredibly skilled at researching, articulating, and then asking questions; two of the three had a significant technical background.
2. I have listened to multiple administrations' Press Secretary broadcasts and was amazed at the low quality of challenging questions from the Press Pool. Never mind the low-quality answers.

You see, I had the privilege of taking journalism in high school with the fantastic teacher, Kelly McWilliams. She and Paul Harvey's video on the topic taught me high integrity journalism. I did not say I was good at it, but I had a fantastic foundation to judge other people calling themselves "journalists."

How often have I screamed in my head or blurted out, "YOU'RE NOT ASKING THE RIGHT QUESTION!"

So, for government security improvements and national security improvements, I thought, let us end an excuse that congressional/parliament leadership might not know what to ask for from the government departments to hold them accountable for their security posture. If Congress/Secretary of/IGs was somewhat the BoD or Audit department of the rest of government (not a political statement), these might be used to challenge them.

I even thought, would it not be cool to have a law requiring these metrics to be reported publicly on a monthly/quarterly basis? I would call it the Joint Interagency Maturing Security Act. You never know … any legislative (local, state, federal, international) member wants to chat; you know where to find me.

So, this list is simple, Congress/Parliament/Secretary/IG requests from the departments or branches that they have oversight, a visualization (not many words) on these metrics, say, quarterly to start, and beginning, say next quarter, or let us be nice, the first quarter of next fiscal year. Maybe post someplace easy to remember. *SecurityMetrics[dot]gov* or *securitymetrics[dot]co[dot]uk*, for example. Get those big poster boards out that we see on C-SPAN behind the legislator desk and let us start moving the needle of government security.

Top 25 Security Metrics for Congress/Secretary of to Ask Government CISOs/CSOs

	The Metrics	Best Score
1	Percentage of technology assets that are EOL/EOS that will not be replaced by the end of the fiscal year	0%
2	Percentage of technology assets that have individual signed off compliance to agency Security Technical Implementation Guides (STIGS)	100%
3	Percentage of buildings with government assets that have been assessed by certified assessors, in last 12 months	100%
4	Percentage of government suppliers with agency security requirements, that have been onsite reviewed by certified assessors, in last 12 months	100%
5	Percentage of agency's assets that are certified, by certified assessors, in the last six months as being fully compliant with regulatory security requirements	100%
6	Percentage of agency technology assets that have latest firmware and software security updates	100%
7	Percentage of breaches reported to agency's investigation authority with 24 hours	100%

	The Metrics	Best Score
8	Percentage of critical infrastructure that has been tabletop and mock event in the last 12 months	100%
9	Percentage of physical security vulnerabilities that will not be fixed by the end of the fiscal year	0%
10	Percentage of fraud vulnerabilities that will not be fixed by the end of the fiscal year	0%
11	Percentage of technology-based security vulnerabilities that will not be fixed by the end of the fiscal year	0%
12	Percentage of employee-based security personnel that are industry trained and certified in their discipline	100%
13	Percentage of government investigators that are trained and certified, by accredited professionals, on insider threat investigations	100%
14	Percentage of agency's departments feed data, indicators of compromise (IOCs), and intel that go into the Agency/Department's Insider Threat SOC	100%
15	Percentage of technology assets that have not been fully assessed for vulnerabilities in the last 12 months	0%

	The Metrics	Best Score
16	Percentage of physical locations that have not been fully assessed for vulnerabilities in the last 12 months	0%
17	Percentage of government investigators that are trained and certified by accredited professionals, on supply chain investigations	100%
18	Percentage of transportation vehicles carrying government assets that have near real time location tracking fed back to the Agencies GIS platform	100%
19	Percentage of agency employees that have had an updated background check in last three years	100%
20	Percentage of agency contractors with up-to-date background checks, per the contract	100%
21	Percentage of records holding personally identifiable information (PII), payment card industry (PCI), customer proprietary network information (CPNI), Protected Critical Infrastructure Information (PCII), etc. that are not encrypted and set to principle of least privilege (PLP)	0%

	The Metrics	Best Score
22	Percentage of government websites and databases with 24x7 espionage monitoring/link analysis by humans	100%
23	Percentage of local congressional offices that have full security coverage (physical, cyber, personnel) to same level as their DC offices/technology/assets	100%
24	Percentage of agency with a yearly report on "State of Security" from their CSO and CISO delivered to Congress/Secretary ____ (and the public?)	100%
25	Percentage of proposed solutions that are currently not funded, that allows for a breached asset to remain vulnerable	0%
	Extra Credit: Percentage on technology assets physically in the Halls of Congress/Secretary HQ office, that have unmitigated vulnerabilities	0%

Interesting Sources of Measurable Threat, Vulnerability, Breach Discovery Information

One Guy's Perspective: Ok, over 180 pages of metrics, I thought I would go off the beaten path, and brain dump some fun sources of threat, vulnerability, breach discovery data for the analytics professionals that got this far. Obviously, collecting this data, determining the indicators to look for, and questions to answer for your specific organization then analyzing it, and reporting it would take a whole other book or class. But when you are bored, grab 10MB of this data and spend a little time in conversation with the data and see what you find.

Interesting Sources of Measurable Threat, Vulnerability, Breach Discovery Information

	Source
1	Purchasing Card data
2	Travel Credit Card data
3	Physical Key Management Logs
4	Badge usage data
5	Virtual Private Network (VPN) Data (Geolocation Info)
6	Ethics Hotline data
7	Click Stream Data of your websites
8	Expense reports
9	International Phone calls inbound/outbound
10	abuse@ and security@ email boxes
11	Door Alarm data
12	Energy usage data
13	Radio Frequency (RF) energy data
14	Third party cameras pointing at your assets
15	Company vehicle onboard computer (OBD II) data
16	Bill of Lading data
17	Customs data (yours, your suppliers, your competitors)
18	Medical Benefits "anomaly"/fraud investigation data
19	Proxy Logs (Size and File Extensions)
20	Visitor Sign in Logs
21	Accounts Payable data

	Source
22	External Requests for organization information (Example: press requests, FOIA, Shareholder requests, subpoenas, kid school project, etc.)
23	LinkedIn Job change data
24	Domain History data
25	Accounts Payable/Accounts Receivable Bank Routing data
	Extra Credit: Customer contract data looking for Brown M&Ms (-type of) requirements

Worst Excuses for Lack of Comprehensive Security Metrics

One Guy's Perspective: Okay, humor time, let us close out with a couple of funny lists, first up, funny excuses because a security organization does not do or like metrics. I am not saying I have heard any of these and I am not saying I have not used any of these. In America, we would call it "Pleading the Fifth."

Top 25 Worst Excuses for Lack of Comprehensive Security Metrics

	The Excuse
1	I/We do not have support from senior leadership to build a comprehensive metrics program
2	I/We have all the resources we need, but do not have time or human resources to build out a metrics program
3	I/We have a metrics program and have not been asked to improve it
4	I/We have a statistics program and have not seen a need for a (full) metrics program
5	I/We are afraid a metrics program will reveal more work that we can handle
6	I/We are afraid a metrics program will reveal failures to our leadership, in our own organization
7	I/We are concerned the metrics may reveal something that requires government or customer notifications
8	None of my peers in the industry told me they have a metrics program
9	I/We do not need more work
10	I/We do not have a skill to do the technical part of a metrics program
11	That IT's job
12	Metrics are only for the business teams and is not needed for administrative functions like security

	The Excuse
13	I do not want to start a metrics program because I am going to retire soon
14	I /We do not mind statistics, but metrics and analytics is too much
15	My boss is not a numbers person
16	We already know what we need to improve, we do not need to measure to improve
17	I/We might lose resources if the metrics show I am underutilizing my resources
18	I'm/We are security people not numbers/data people
19	I/We cannot afford any of the metrics/dashboard tools
20	Our tools can produce reports when we need them
21	I/We used to have a metrics program and it was a disaster
22	I/We do not want to measure areas we already know are bad
23	There are no industry standards for security metrics
24	Metrics programs are only for the IT Security/Cyber Security teams, not for the other security functions/physical security
25	We do not know where the data is
	Extra Credit: We don't know who NIST is

Worst Requests for Security Metrics

One Guy's Perspective: Okay, humor time again; let us close out with a final funny list. This one was started in a post by my professional peer, Tony Cifaretto, one of the leading technology investigative forensics guys I know. This list is about crazy requests from leadership and others related to various security metrics and stats. I may have even been in situations where I produced weekly/monthly/near real-time metrics and leadership admitted they never looked at them. Again, I am not saying I have heard any of these, and I am not saying I have not asked for any of these. In America, we would call it "Pleading the Fifth."

Top 25 Worst Requests for Security Metrics

	The Request
1	What do we need in the budget to mitigate these threats next year?
2	How many probes/attacks did we get today?
3	How many forensic examinations were successful?
4	What is your KPI for successful forensic examinations?
5	What do we need in the budget to mitigate all of our vulnerabilities? ("Stop Hiring Humans")
6	How many cameras do we have?
7	How many badge readers do we have?
8	How many computers do not have anti-virus software?
9	How much money have we recovered from fraud investigations this year?
10	How much money have we recovered from loss prevention investigations this year?
11	Customer is asking how many attacks have there been on their data?
12	Consumer is asking how many people have access to their information? (Think across the Open Systems Interconnection (OSI) stack.)
13	Who has access to our internet website?
14	How people made unauthorized purchases on the company credit card?

	The Request
15	How many personnel used our computers for personal use?
16	How many people use their company telephone for personal use?
17	How many minutes late was Billy Bob according to his badge usage?
18	How many breaches have we had this year? And how many virus attacks have we had this year? If the virus number is higher than breaches ... you have a serious problem with definitions.
19	How many holes are in our chain-link border fence?
20	How many pieces of mail did we scan enterprise-wide this year?
21	How many people entered our buildings so far this year?
22	How many vulnerabilities do we need to fix?
23	How much fraud involved one of our systems or applications?
24	How many investigations have we asked our (non-security) employees to work on?
25	What would be the measurable impact if we ended all of our guards?
	Extra Credit: What is the 5-year projection plan on security ____________?

Contributing to the 2024 Edition

I was a little selfish not to ask 1001 people in my impressive and grace-filled network for their input on the 2023 Edition. I wanted to set a baseline, a low bar, but still, a baseline, to reduce the excuses on the security and business/government sides. I hope people can use this tool in 2023 and justify remarkable improvements from its results in 2023 and beyond.

For any of the metrics in the book, please contact me if you have some recommended changes for 2024, for example, new metrics, replacement, or even simple reworded metrics. I would even love some stories of positive and negative reactions. I suspect there might even be one person that mysteriously orders this book for their senior leadership as an anonymous *gift*.

Here is a list of possible 2024 topics that I already have in mind to expand the book/tool. For some of them, I will reach out to far more competent experts for input; for others, I have some ideas. Again, if you want to see a topic or function covered, individual metrics, or, why not, your own *Top 25* for a new topic, send it; I will be glad to give you the honor you deserve.

- Threat Management
- Intelligence Management
- Industry-specific
    - Faith-Based Community
    - Law Enforcement
- Cyber Security (this is hard due to its broadness, definition, and scope conflicts in the industry)
    - Software Development (Coding, Software bill of materials (SBOM))
    - Access Administration
- After Action Management
- Active Threat Preparation and Response
- Acceptable Use Policy (AUP) Monitoring
- Location (building/country) Risk Assessment (not percentage)
- Supply Chain
    - Transportation

Closing Arguments

The sets of metrics have been formed, hardened, and packaged in this short book. We hope in future years we will expand them into other challenging areas. We hope this will excite some, cause laughter to others, challenge others to challenge others, scare others, and even cause some anger, but in the end, if it causes the security needle to move one way or the other, we succeeded. This is not meant to be scientific, academic, or even *proper* in every way; honestly, it's intended to eliminate excuses and improve security "1% a day."

Your organization needs this book … this tool; it is not a law or a regulation, but a guide. You cannot throw an excuse for inaction that would stick. I even collected a few excuses in one of the last chapters that you are not allowed to use anymore as they have already been taken. I'm joking. I know politics are involved so excuses come into play.

This is my first book; it has been a blast and energizing to write. It has triggered three more book ideas that I hope I am blessed to be able to write soon. I hope that future editions are enhanced with input from readers and by all organizations that desire to challenge their people, culture, and organization. May future editions be equally valuable to you as an individual and to your organizations.

May the Lord Bless You in this journey.

Jim McConnell
Just One Guy's Perspective
#NeverBored #NoMoreExcuses #ResumeBucketList

Definitions

If you and your organization use different terms or scope, that is ok; just think about your wording with my scope or whether your scope of that term is broad enough.

Term	My Perspective Definition
ALL	*All* in every language is All – All business units, all employees, all contractors, all suppliers, all countries, all customers, etc.
Analytics	The evaluation of information or data to ask and answer questions that drive changes in an organization.
Assessment	Checklist (egad), tools, the pen test, tabletop, etc.) – finding and hopefully exploiting vulnerabilities with actionable solutions.
Asset	Hard/physical asset (e.g., building), electronic asset (e.g., stuff on a computer), intellectual property, processes, people.
Assets	Physical, Electronic, and Intellectual Assets that have a material impact/risk to the organization, suppliers, companies, and societies.

Term	My Perspective Definition
Audit	The (independent) verification and validation of a control to verify it meets the legal, regulatory, or business need to protect the business, government, or consumer.
Authorized Fraud	The commitment of fraud by a person that is technically approved/authorized to perform a specific function. They just abuse that authority to commit the fraud.
Background Investigations	Investigations that are generally PROACTIVE on the review of the background of a person (background checks/investigations) or company (due diligence) to determine the viability and integrity of the individual or company before a relationship is established or renewed.
Buildings	Locations where an organization's assets are located, this can be as simple as a cinder-blocked building, a supplier's warehouse, all the way to interesting "buildings" like "street furniture" (an old Telco term).

Term	My Perspective Definition
Classified information vs. information classification	Classically speaking, *classified information* has been tied to sensitive and critical information assets that need protection for national security interests. Whereas *information classification* is, the establishment of information owner-defined labels and controls for an organization's non-public information. These terms have been used interchangeably, but, classically, *classified information* is, actually, the implementation of *information classification* in a government/national security environment.
Compliant	A term that indicates an assessed element (organization, process, technology, function) meets the requirements of a documented standard, generally accepted set of requirements, law, regulation, or legally binding document like a contract. *Compliant* does not mean the element is secure.
Cyber Security	The functional role, hopefully in a converged security program, to "Secure stuff with wires" from breaches in confidentiality, integrity, and availability.

Term	My Perspective Definition
Distributor/ Distribution	A middle entity or person in the supply chain of a product somewhere between manufacturer and end user.
Enterprise-wide	A way of describing the scope of an element of a security program/function/project where it represents all organization elements under the Chief Executive Officer.
Ethical Stop	A term I coined many years ago in my fraud and insider threat presentations and classes; the goal of a security and ethics department to encourage people to not go beyond a certain point of unethical behavior. For instance, stop at stealing pens …
First Responder	An individual trained in one or more emergency response disciplines that is generally (part of) the first individuals on the scene to care for others or prevent further crisis.

Term	My Perspective Definition
Fraud	Any activity that relies on deception in order to achieve a gain. Fraud becomes a crime (thus a security concern) when it is a *knowing misrepresentation of the truth or concealment of a material fact to induce another to act to his or her detriment* based on the Fraud Triangle.
Fraud Investigations	Sometimes called a *Fraud Examination*; investigations that specifically focus on fraud and usually involve specialized skill sets around financial crimes, occupational fraud, and fraud analytics. (I hope to do a *Top 25* just in this area in the next edition.)
Fraud Management	The end-to-end coverage of activities involved in preventing, discovering, investigating, and mitigating fraud, including but not limited to fraud awareness, pro-active fraud analytics, pre-deploy fraud assessments, mitigation, or corrective active project management … oh, and investigations.

Term	My Perspective Definition
HR Investigations	Personnel investigations that involved employees and usually, but not always, are run by HR team members and not the security department. These investigations tend to involve, what some people might call social policy violations, like hostile work environments, and sexual harassment.
Incident Response	This is the immediate reaction to some type of safety or security incident and involves steps to reestablish a safe environment for ongoing business operations. There may be one or more investigations after the priority of response is completed to figure out the cause and ongoing mitigation that triggered the incident.
Information	Classic definition: facts provided or learned about something or someone. Contextual definition: facts supplied or learned about an organization's assets or assets it is entrusted to manage (and secure) regardless of form or where it is located.

Term	My Perspective Definition
Information Security	The prevention, detection, and corrective/response actions to protect the confidentiality, integrity, and availability of an organization's information or information entrusted to the organization.
Information Security vs. (Cyber, Application, Network, Privacy, etc. Security)	It has been said that *cyber security* was once defined as *secure things with wires*. The securing of electronic stuff tends to incorporate not ***just*** security of information, but also the security of services operating within the technology of *cyber, applications*, and *networks*. Security of both information and these other areas are important, but without information, organizations do not exist; without technology-based services, organizations cannot run.
Infrastructure	I like to think of infrastructure as all that technology you do not see, but unplug it, do not maintain it, and your security department and security controls go downhill fast.

Term	My Perspective Definition
Insider Threat	The pre-action awareness and indicators of an insider in an organization is planning to commit a security offense in the near future. Insider Threat has also been used as the label for the prevention, detection, and response to an insider's malicious actions.
Insider Threat (Non-Malicious)	A term sometimes used to show that an insider is seen/used to carry out a malicious act, but the insider is simply following standard protocols or weak controls. Another malicious party (insider or outsider), generally has social-engineered this insider to help malicious party.
Investigations	The collection, analysis, and reporting of the facts to prove or disprove an allegation of wrongdoing or incident. Facts can include, but are not limited to, any evidence detectable or performed by the five senses. Investigations can include areas such as a safety investigation (e.g., derailment of a train) that is not necessarily a security investigation. Investigations, in this *Top 25*'s context are not limited to any specific group of human victims or perpetrators.

Term	My Perspective Definition
Law Enforcement Agency	An entity commissioned by the laws of the land to prevent, detect, respond to crime committed against individuals, companies, governments, or property.
Link Analysis	An analytical investigative technology that is usually presented visually on a board or with software, to connect elements, entities, evidence of an incident, background check, or due diligence to better visualize interconnections and possible related incidents.
Loss Prevention	Usually tied to the retail, transportation, logistics industry, with focus on preventing and investigating the thief of products throughout the supply chain.
Pen Test	A structured (hopefully) vulnerability exploitation process to test prevention, detection, and response controls of a target. This target is usually a technology-based target or physical building/area-based target. Using and exploiting people, process, technology, tools, and buildings to accomplish this scope of the test.

Term	My Perspective Definition
Personnel	Employees and non-employees (e.g., suppliers or contractors) with physical access or logical/electronic access to your (confidential) assets. Generally, this does not include the public or customers.
Personnel Investigations	Specific investigation area that is focused on allegations involving employees. Some people will also include contractors in this area.
Risk	The (hopefully measured) probability of *something bad is going to happen*.
Risk Management	The management of controls to reduce the probability that *something bad is going to happen* to the lowest level practical for an organization, taking into considering brand, legal, cost, operations, productivity, etc.
Security	The prevention and detection of, and response to a crime or violation of company policy.
Security Assessment	The assessment of the defined scope to determine if the assets of that scope are susceptible to (or even a victim of) being part of a crime or violation of an organization's policies.

Term	My Perspective Definition
Security Audit	The assessment of the defined scope to determine if the assets of that scope are susceptible to (or even a victim of) being part of a crime or violation of an organization's policies. Security Audits are also tied to compliance scope and usually reported to a board, audit committee, or other oversight organization.
Security Department Technology	The tools used or managed by the Security Department such as websites, databases, analytics tools, desktops, servers, IoT, applications, network, mobile devices, or software (open source, commercial, proprietary).
Security Investigations	An investigation that involves a crime or violation of an organization's policies/Code of Conduct.
Security Response	The tools, techniques, and procedures used to prevent, detect, and respond to an incident (observe–orient–decide–act (OODA) Loop anyone).
Software	Firmware, disk-based software, cloud-based software, operating system level, database level, application level, etc.

Term	My Perspective Definition
Subcontractors	Any external entity, which is used by your direct suppliers, from which you get products or services, that affect the risk to the company, regardless of whether you pay them directly or indirectly, and regardless of whether you have them under contract.
Supply Chain (aka vendor, supplier, third-party)	Any external entity that you get products or services from, that affect the risk to the company, regardless of whether you pay them directly or indirectly, and regardless of whether you have them under contract. (More detailed discussion on "What is a supplier?" is posted in my documents on LinkedIn.)
Supply Chain Investigations	Investigations that involve a supplier or vendor and the organization. Like other specialized investigations, this one also takes special skill sets to cover the unique attributes related to the supply chain.
Threat	The direct or implied communication of intent to inflict harm or loss on another person, entity.

Term	My Perspective Definition
Threat Assessment	The one-time or recurring evaluation of actions, indicators, or communications by a person, entity, or technology that is believed to be planning a breach, attack or other malicious action against another person, technology, or entity.
Threat Management	The ongoing actions to prevent, detect, and respond to threats by an individual, entity, or technology.
Training	The education of an individual through any organization's acceptable means, which is trackable, culturally adapted, that communicates, shows implementation, and tests students on the security, rules, standards, guidelines, good practices, procedures recommended and required for an organization's safe and secure success. This does not mean an organization's employee is the trainer/teacher, but I do recommend a dedicated employee manage the training program.

Term	My Perspective Definition
Training Sustainment Timing	Some training should be yearly, and some monthly; some training should be recurring via, say, a screen saver splash screen. Some should be repeating the same training, others should get naturally more challenging than the last class.
Vulnerability	Weakness in a physical or electronic asset, procedure, or implementation, which could be exploited or triggered by a threat source or operational security assessor. I understand that some people, cultures, environments, or organizations do not like the word *vulnerability*, so they will soften it with the phrase *security gap*. I am not going to go there, I respect the organization's culture, political, and optics challenges, but I'm going to stick with *vulnerability*.

Term	My Perspective Definition
Vulnerability Assessment	The discovery or detection of weaknesses in people, processes, technology, or other company assets (e.g., buildings) that would make them susceptible to being used to commit a crime or violation of organizational policies. This assessment could also involve probability, mitigation, and impact.
Vulnerability Discovery	The malicious or operational security method for finding a vulnerability in a person, physical asset, electronic asset, procedure, or implementation. *Discovery* does not necessarily mean the vulnerability was exploited. For example, I see a door with poor outside hinges or I find a computer running EOL/EOS software, but in either case, I do not take advantage of these findings and compromise the asset.
Vulnerability Exploit(ed)	The active application of methods that uses the discovered weakness to take advantage of the weakness to compromise an asset by the threat source or operational security assessor.

Term	My Perspective Definition
Vulnerability Management	The ongoing actions to prevent, detect, and respond to threats by an individual, entity, or technology. Some may consider this patch management, but patch management is just a subset of overall vulnerability management.
Wholesale	A middle entity or person in the supply chain of a product somewhere between manufacturer and end user. Usually, the first entity after the manufacturer that sells to various parties, but not usually a retail consumer.

Made in United States
Cleveland, OH
04 November 2026

47642874R00123